Beverage Sensory Modification

Beverage Sensory Modification

Special Issue Editor

Manuel Malfeito Ferreira

MDPI • Basel • Beijing • Wuhan • Barcelona • Belgrade

Special Issue Editor
Manuel Malfeito Ferreira
University of Lisbon,
Portugal

Editorial Office
MDPI
St. Alban-Anlage 66
4052 Basel, Switzerland

This is a reprint of articles from the Special Issue published online in the open access journal *Beverages* (ISSN 2306-5710) from 2017 to 2019 (available at: https://www.mdpi.com/journal/beverages/special_issues/sensory_modification)

For citation purposes, cite each article independently as indicated on the article page online and as indicated below:

LastName, A.A.; LastName, B.B.; LastName, C.C. Article Title. *Journal Name* **Year**, *Article Number*, Page Range.

ISBN 978-3-03921-393-1 (Pbk)
ISBN 978-3-03921-394-8 (PDF)

Contents

About the Special Issue Editor

Manuel Malfeito-Ferreira is Professor at the Instituto Superior de Agronomia (ISA), University of Lisbon. His research is mainly focused on food and wine spoilage yeasts, especially concerning volatile phenol production by Brettanomyces bruxellensis. His more recent projects include consumer studies on wine acceptance and the development of a new tasting approach based on emotional responses. He is frequently invited to speak in technical seminars about wine microbial stability and wine tasting all over the world.

Editorial

Beverage Sensory Modification

Manuel Malfeito-Ferreira

Linking Landscape, Environment, Agriculture and Food (LEAF) Research Center, Instituto Superior de Agronomia (ISA), University of Lisbon, 1349-017 Lisboa, Portugal; mmalfeito@isa.ulisboa.pt

Received: 22 May 2019; Accepted: 25 June 2019; Published: 3 July 2019

The Special Issue on "Beverage Sensory Modification" gathers a series of articles that feature the broad sense of sensory modification, either by improving flavor, taste, and mouthfeel properties or by preventing their spoilage. The scope goes further than the usual technological measures that modulate sensory properties and includes the psychological and cross-modal influences, where the sensory modification is on the subject's brain and not on the object's physical-chemical properties.

The beverage industry usually addresses the question of modifying sensory characteristics by meeting the innate attraction for sweetness either by adding different sweetening agents, different aroma molecules, or changing dishware forms, which are known to increase the in-mouth sweet perception. Non-nutritive sweeteners have been used as substitutes for nutritive sweeteners with the goal of preventing obesity and dental caries. The main factor responsible for the difference in taste between beverages containing a nutritive sweetener and those containing a non-nutritive sweetener is the temporal profile of sensory attributes. However, Gotow et al. [1] demonstrated that this difference was only observed in water and not in coffee beverages, probably because of flavor properties that masked the sweetener effect. The cross-modal flavor–taste interactions also revealed the importance of the matrix effect as described by Wang et al. [2] using skim milk. These authors showed that a vanilla aroma did enhance the perceived sweetness while this enhancing effect was lower than that of sucrose on the vanilla flavor. The cross-modal interactions also include visual and taste senses. In particular, research indicates that roundness (as opposed to angularity) is consistently associated with an increased sweetness perception. However, Machiels [3] did not confirm these results using two different cup forms (round versus angular) with a butter milk drink and a mate-based soft drink. Interestingly, a correspondence was found between the angular cup and a more bitter taste only for the soft drink. The sweetener sucrose may also be used for other purposes than increasing sweetness [4]. These authors showed that it also affects the texture and creaminess of a new product based on partially demineralized sweet whey and gelatin added to milk powder and cassava starch. Creaminess and firmness were also promoted by the cassava starch. Overall, these four articles highlight that food or beverage matrixes exert a significant effect on taste and mouth-feel studies and are indispensable to validate preliminary assessments using water solutions.

Wines are also a frequent object of sensory studies, gathering researchers with different scientific backgrounds. The shape and size of the wine glass was shown to affect the different wine aromas in the headspace [5] Moreover, Spence and Wang [6] demonstrated that the quality of the wine was rated as higher and the celebratory mood of the participant was also higher following the sound of the cork pop when compared with a screw-cap opening. The cross-modal interaction received here of another input from the senses of hearing, smell, and taste influences hedonic responses. Under a different scope, off-flavors also deserve the attention of researchers. For instance, the world-famous Chardonnay from Burgundy may be affected by oxidative notes that indicate premature aging [7] The highly debated "horse sweat" taint was also reviewed, encompassing technical preventive measures and the influence of volatile phenols on sensory attributes [8].

This special issue enables consumers to be aware of the work that is being carried out by leading research teams in areas that may be regarded as case studies for the whole of the food and beverage industries.

References

1. Gotow, N.; Esumi, S.; Kubota, H.; Kobayakawa, T. Comparison of Temporal Profiles among Sucrose, Sucralose, and Acesulfame Potassium after Swallowing Sweetened Coffee Beverages and Sweetened Water Solutions. *Beverages* **2018**, *4*, 28. [CrossRef]
2. Wang, G.; Hayes, J.; Ziegler, G.; Roberts, R.; Hopfer, H. Dose-Response Relationships for Vanilla Flavor and Sucrose in Skim Milk: Evidence of Synergy. *Beverages* **2018**, *4*, 73. [CrossRef]
3. Machiels, C. Bittersweet Findings: Round Cups Fail to Induce Sweeter Taste. *Beverages* **2018**, *4*, 12. [CrossRef]
4. Miraballes, M.; Hodos, N.; Gámbaro, A. Application of a Pivot Profile Variant Using CATA Questions in the Development of a Whey-Based Fermented Beverage. *Beverages* **2018**, *4*, 11. [CrossRef]
5. Parpinello, G.; Matteo, M.; Arianna, R.; Andrea, V. Effect of Different Glass Shapes and Size on the Time Course of Dissolved Oxygen in Wines during Simulated Tasting. *Beverages* **2018**, *4*, 3. [CrossRef]
6. Spence, C.; Wang, Q. Assessing the Impact of Closure Type on Wine Ratings and Mood. *Beverages* **2017**, *3*, 52. [CrossRef]
7. Ballester, J.; Magne, M.; Julien, P.; Noret, L.; Nikolantonaki, M.; Coelho, C.; Gougeon, R. Sensory Impact of Polyphenolic Composition on the Oxidative Notes of Chardonnay Wines. *Beverages* **2018**, *4*, 19. [CrossRef]
8. Malfeito-Ferreira, M. Two Decades of "Horse Sweat" Taint and Brettanomyces Yeasts in Wine: Where do We Stand Now? *Beverages* **2018**, *4*, 32. [CrossRef]

Article

Comparison of Temporal Profiles among Sucrose, Sucralose, and Acesulfame Potassium after Swallowing Sweetened Coffee Beverages and Sweetened Water Solutions

Naomi Gotow [1], Shinji Esumi [2], Hirofumi Kubota [2] and Tatsu Kobayakawa [1,*]

[1] Human Informatics Research Institute, National Institute of Advanced Industrial Science and Technology (AIST), Tsukuba Central 6, 1-1-1 Higashi, Tsukuba, Ibaraki 305-8566, Japan; nao-gotow@aist.go.jp

[2] Products Research & Development Laboratory, Research & Development Headquarters, Asahi Soft Drink Co. Ltd., 1-1-21 Midori, Moriya, Ibaraki 302-0106, Japan; shinji.esumi@asahiinryo.co.jp (S.E.); hirofumi.kubota@asahiinryo.co.jp (H.K.)

* Correspondence: kobayakawa-tatsu@aist.go.jp; Tel.: +81-29-861-6730

Academic Editor: Manuel Malfeito Ferreira

Received: 5 December 2017; Accepted: 28 March 2018; Published: 2 April 2018

Abstract: Non-nutritive sweeteners have been used as substitutes for nutritive sweeteners with the goal of preventing obesity and dental caries. The main factor responsible for the difference in taste between beverages containing a nutritive sweetener and those containing a non-nutritive sweetener is the temporal profile of sensory attributes. In this study, untrained panelists performed a time–intensity evaluation of sweetness, using one coffee beverage containing a nutritive sweetener (sucrose) and two coffee beverages containing non-nutritive sweeteners (sucralose or acesulfame potassium (acesulfame K)). They evaluated continuously perceived intensity of sweetness for 150 s after swallowing each coffee beverage. We did not detect a significant difference in temporal profiles among the three coffee beverages. To investigate why the temporal profiles of the three coffee beverages followed similar traces, all untrained participants who had participated in the coffee beverage session also performed a time–intensity evaluation of sweetness using three water solutions (sucrose-sweetened, sucralose-sweetened, and acesulfame K–sweetened deionized water). We observed a significant difference in temporal profiles among the three water solutions. These results indicate that differences in the temporal profiles of coffee beverages might be masked by factors other than the sweetness of the sweetener.

Keywords: temporal profile; time–intensity evaluation; sweetener; coffee beverage; water solution; untrained panelist

---

1. Introduction

Sucrose, a disaccharide, is one of the most common nutritive sweeteners used today, and provides metabolizable energy [1]. Overconsumption of sucrose causes chronic diseases such as obesity, metabolic syndrome leading to diabetes, and cardiovascular diseases [2]. In order to prevent these illnesses, non-nutritive sweeteners with very low or no calories have been used as substitutes for nutritive sweeteners [3,4].

Currently, the carbohydrate content of ordinary products such as coffee beverages is around 7.5 g per 100 mL [5] in Japan. In order to use the term "low-sugar", it is necessary to decrease the carbohydrate content to less than 2.5 g per 100 mL of beverage [6], or to reduce the carbohydrate content by more than 2.5 g per 100 mL relative to the content in ordinary products [5]. In the European

Union, a "low-sugar" product has been defined as one that contains less than 5 g of sugar per 100 g for solids or 2.5 g of sugar per 100 mL for liquids [7]. The United States Food and Drug Administration has not defined the term "low-sugar" [8]. Non-nutritive sweeteners such as sucralose and acesulfame potassium (acesulfame K) are mainly used to enhance the sweetness of "low-sugar" products that contain fewer carbohydrates than conventional products. Per unit mass, sucralose and acesulfame K are several hundred times sweeter than sucrose [9,10].

The temporal profile of sensory attributes is the factor most responsible for a taste difference between a product sweetened with sucrose and one sweetened with a non-nutritive sweetener [11]. The temporal profiles of currently used non-nutritive sweeteners are not consistent with that of sucrose [12]. When investigating the temporal profiles of sensory attributes (i.e., sweetness) of water solutions and beverages, the measurement method used most frequently is a time–intensity evaluation [13,14]. This method records how the perceived intensity of sensory attributes changes over time [15]. To compare the temporal profiles of sweetness among multiple sweeteners, time–intensity evaluations are performed using water solutions [16–20] or foods (i.e., espresso coffee [21], chocolate milk [22], mixed fruit jam [23], milk chocolate [24], and snacks [25]). Melo and colleagues [26,27] sought to develop diabetic chocolate perceptually similar to ordinary chocolate in terms of its temporal profile. Trained panelists performed a time–intensity evaluation of sweetness using ordinary chocolate containing sucrose, as well as diabetic chocolates containing sucralose or stevioside. Their results revealed that the temporal profiles of sweetness were similar between ordinary and diabetic chocolates. Thus, time–intensity evaluation is considered to be a useful method for developing new products with temporal profiles that are as similar as possible to those of ordinary products.

The purpose of this study was to investigate whether the temporal profiles of sweetness differed between an ordinary coffee beverage containing nutritive sweetener (sucrose) and low-sugar coffee beverages containing non-nutritive sweeteners (sucralose or acesulfame K). Participants also performed a time–intensity evaluation of sweetness using sucrose-sweetened, sucralose-sweetened, and acesulfame K–sweetened water solutions as control samples. Gotow and colleagues [28] developed a time–intensity evaluation system for untrained panelists, which was used in this study.

2. Materials and Methods

2.1. Participants

This study was conducted in accordance with the revised version of the Declaration of Helsinki. All procedures in this study were approved by the ethical committee for ergonomic experiments of the National Institute of Advanced Industrial Science and Technology, Japan. Informed written consent was acquired from all participants. Ninety-four volunteers (39 female and 55 male) between the ages of 20 and 29 years (mean age $\pm$ standard deviation = 22.29 $\pm$ 1.90 years) participated in the experiments; volunteers contacted us after seeing our recruitment announcement on a website for the local community.

2.2. Materials

The experiments used eight types of non-released canned samples (Asahi Soft Drink, Tokyo, Japan): four sweetened coffee beverages and four sweetened water solutions. Sucrose, sucralose, and acesulfame K were used as sweeteners for the test trials, and stevia was used as the sweetener for the exercise trial.

Sucralose, acesulfame K, and stevia are 600, 200, and 350 times sweeter per unit mass than sucrose, respectively; here, the sweetness of sucrose was defined as 1 [9,10,29]. Most low-sugar canned coffee beverages available in Japan contain milk and have sweetness levels of 40–55. Therefore, we prepared a coffee beverage with milk that had a sweetness level of 50. More specifically, we calculated the concentrations satisfying the following equation: [sample concentration (g/L)] $\times$ [sweetness equivalence ratio of each sweetener, relative to sucrose] = 50. Accordingly, we prepared samples sweetened with 5% sucrose, 8.33×10^{-3}% sucralose, 2.50×10^{-2}% acesulfame K, and 14.28×10^{-3}% stevia.

For coffee beverages, we extracted components from 56.5 g of coffee beans (a blend of several medium-dark roasted Arabica beans), ground to a specified size, with deionized water at approximately 95 °C; added sweetener and 120 g of milk (3.6% fat, sterilization at 130 °C, Ohayo Dairy Products, Okayama, Japan) to the coffee extract; and then diluted the mixture in a measuring cylinder to 1 L with deionized water. After canning, each sample was subjected to retort sterilization. For water solutions, we canned deionized water containing each sweetener, and then performed retort sterilization. The canning process was completed using industrial equipment at the Asahi Soft Drink Co. Ltd.

We opened each package of canned sample, salt-free cracker ("Levain Classical non-salt topping", Yamazaki-Biscuits, Tokyo, Japan), and mineral water ("Asahi oishii mizu Fujisan", Asahi Soft Drink, Tokyo, Japan) 1 h before the start of the experiment. The salt-free cracker and mineral water were used to clean the participant's oral cavity [30,31]. A coffee sample or water solution (10 mL) was measured using a macropipette and poured into a paper cup (capacity 90 mL, Part Number SM-90-3, Tokan Kogyo, Tokyo, Japan). We poured one type of sample into each cup. In the interest of food sanitation and aroma retention, we covered the paper cup with a lid (Part Number SM-205-F, Tokan Kogyo), which was removed immediately before the sample was presented to the participants. The salt-free cracker was cut to a size of 2 cm × 2 cm, and one piece of cracker was served in a paper candy cup. Mineral water (10 mL) was measured using a macropipette and poured into a paper cup (capacity 90 mL, Part Number SM-90-3, Tokan Kogyo). Coffee beverages, water solutions, and mineral water were presented at room temperature (approximately 24 °C).

2.3. Time–Intensity Evaluation System

An outline of the time–intensity evaluation system is shown in Figure 1. The structure of this system was detailed by Gotow and colleagues [28,32]. However, to improve the portability of this system, we changed how the voltage output from the load cell was processed. More specifically, after the output voltage was amplified, it was subjected to an analog-to-digital (A/D) conversion at a frequency of 1 kHz on a microprocessor (Part Number Arduino Uno SMD Rev3, Arduino S.r.l., Ivrea, Italy). Average voltage was calculated in time windows of 50 milliseconds, and output to a personal computer via a serial port as a digital value corresponding to a six-point magnitude scale of perceived intensity.

2.4. Procedure

Time–intensity evaluation was performed by one participant at a time in a small room (width 165 cm × depth 275 cm × height 240 cm) shielded from outside noise. The door of the room was closed during measurement. A video camera and intercom were placed inside the room so that the experimenter could monitor and communicate with participants from outside the room. An air cleaner located inside the room was operated continuously to prevent odor retention.

The order of coffee beverage and water solution sessions was alternated between participants. To ensure the reliability and stability of the time–intensity evaluation [33,34], we presented a sample containing stevia in the first trial of each session, which was considered as a training trial. The presentation order of samples containing sucrose, sucralose, or acesulfame K in each session was counterbalanced among participants. The participant rested for approximately 7 min between trials in each session, as well as between the first and second sessions.

All instructions were displayed on the liquid crystal display (LCD) monitor placed in front of the participant. In the first trial of the first session, the participant performed the evaluation according to instructions displayed on the screen, while receiving oral instructions from experimenter. On the first screen of sequential trials, we instructed the participant to evaluate the continuously perceived intensity of sweetness on the tongue after swallowing the samples. On the same screen, based on previous studies [35,36] in which participants reported which part of their anatomy they used to perceived specific sensory attributes (e.g., some participants replied that they perceived vanilla aroma in their mouth), we instructed the participant regarding the part of the anatomy to which they should direct their attention (i.e., "on the tongue"), using an illustration of the sagittal plane of the head with the name of the part labeled (a display is shown in Figure 1). Next, the participant placed a cracker into their mouth to clean the oral cavity and continued masticating it for 15 s before the screen was switched. The participant swallowed the cracker remaining in their oral cavity, and then held 10 mL of mineral water in their mouth. After they transferred the water into the oral cavity, they swallowed it. The participant took 5 mL of sample in their mouth, which they held without swallowing, and then placed the index finger of their right hand into the ring of the time–intensity evaluation system. A countdown was displayed three seconds before the screen showed visual feedback about perceived intensity. The participant swallowed the sample in their mouth at the same time that the countdown reached 0 s and started the time–intensity evaluation of sweetness.

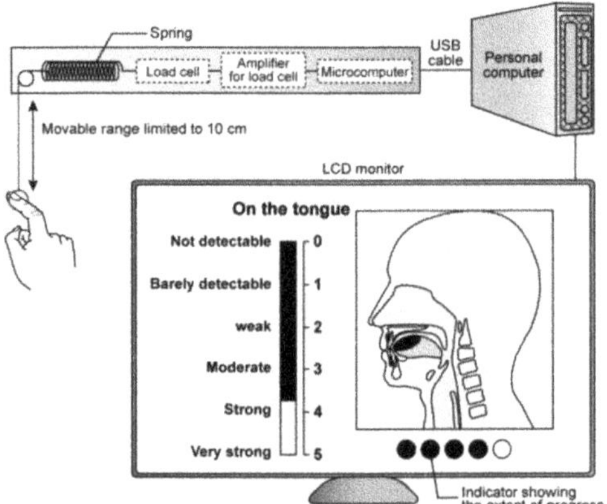

Figure 1. Outline of the time–intensity evaluation system. Participants evaluated perceived intensity by operating a pull-ring, which was a component of the evaluation system. The movable range of the ring was limited to 10 cm by a stopper. Positional information of the ring, synonymous with spring tension, was measured by a load cell, with output expressed as voltage. After the output voltage was amplified, it was subjected to analog-to-digital (A/D) conversion on a microprocessor at a frequency of 1 kHz. Average voltage was calculated in time windows of 50 milliseconds, and then output to a personal computer via a serial port as a digital value corresponding to a six-point magnitude scale of perceived intensity. To provide visual feedback in real time, the value of perceived intensity was displayed on a liquid crystal display (LCD) monitor as a black bar on a six-point magnitude scale (0: "not detectable" to 5: "very strong"). Furthermore, to inform the participant of the time remaining in the evaluation, an indicator of the extent of progress was shown on the screen.

For each sample, participants evaluated perceived intensity over 150 s. We instructed each participant to express the perceived intensity of sweetness by freely operating the pull-ring component of the evaluation system. Participants were not told the length of the evaluation time (i.e., how long they were to evaluate perceived intensity). Instead, to inform the participant of the time remaining in each trial, the screen displayed an indicator showing the extent of progress.

2.5. Analysis

In this study, we regarded the time when the screen was switched to visual feedback of perceived intensity as the starting point of the time–intensity evaluation (i.e., 0 s). We divided the evaluation period from 0 s to 150 s into 75 windows of 2 s each, and calculated the average perceived intensity in each time window. We conducted statistical analysis using these average values.

To investigate whether temporal profiles of sweetness differed among the three types of samples (sucrose-sweetened, sucrose-sweetened, and acesulfame K–sweetened samples), two-way repeated measures analysis of variance (ANOVA) for each solvent session (coffee beverage and water solution sessions) was performed for average values of perceived intensity, with sweetener and time as within-subject factors. Simple effects tests were conducted based on the significance of results obtained with ANOVA. Additionally, when temporal profiles of sweetness did not differ among three types of coffee beverages, we performed two-way repeated measures ANOVA for each sweetener for the average values of perceived intensity, with solvent session and time as within-subject factors.

We used IBM SPSS Statistics 23 (IBM Japan, Tokyo, Japan) for statistical analysis, and considered p values less than 0.05 as statistically significant.

3. Results

3.1. Comparison of Temporal Profiles of Sweetness among Samples in Each Solvent Session

3.1.1. Sweetened Coffee Beverage

The temporal profiles of sweetness of each sweetened coffee beverage are shown in Figure 2. Two-way repeated measures ANOVA revealed a significant main effect of time (F [74, 6882] = 326.78, $p < 0.001$).

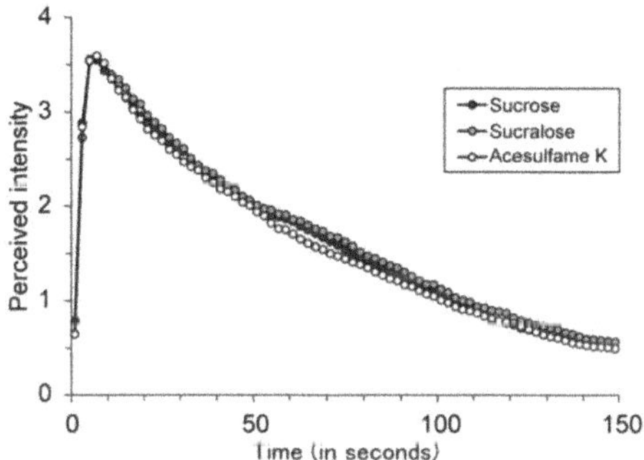

Figure 2. Temporal profiles of sweetness of each sweetened coffee beverage. Temporal profiles of sweetness were obtained for 150 s after participants swallowed sweetened coffee beverages.

3.1.2. Sweetener in Water Solution

The temporal profiles of sweetness of each sweetener in water solution are shown in Figure 3. Two-way repeated measures ANOVA revealed significant main effects of sweetener (F [2, 186] = 10.81, $p < 0.001$) and time (F [74, 6882] = 304.68, $p < 0.001$), and a significant interaction between sweetener and time (F [148, 13764) = 4.74, $p < 0.001$). The results of simple effects tests for interaction revealed significant simple main effects of sweetener in 74 time windows (medians of each time window = 3–149 s), and significant simple main effects of time for all sweeteners ($p < 0.05$). Multiple comparisons of the significant simple main effects of sweetener in those time windows, performed using the Ryan method, revealed significant differences between sucrose and sucralose in 63 time windows (medians = 25–149 s), between sucrose and acesulfame K in 25 time windows (medians = 3–15 s, 71–77 s, and 81–107 s), and between sucralose and acesulfame K in 40 time windows (medians = 3–81 s) ($p < 0.05$; for further details, see in Table 1). These results indicated that in some time windows, the perceived intensity of sweetness was significantly higher for sucralose-sweetened water solution than for sucrose-sweetened water solution, for acesulfame K–sweetened water solution than for sucrose-sweetened water solution, for sucrose-sweetened water solution than for acesulfame K–sweetened water solution, or for sucralose-sweetened water solution than for acesulfame K–sweetened water solution.

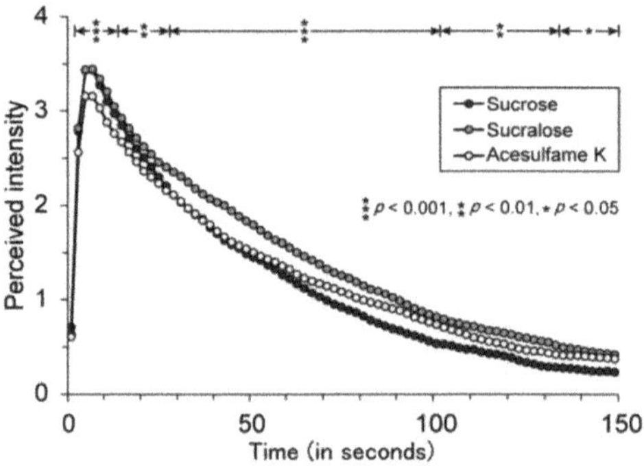

Figure 3. Temporal profiles of sweetness of each sweetener in water solution. Temporal profiles of sweetness were obtained for 150 s after participants swallowed sweetened water solutions.

Table 1. Results of sweetener in water solution: multiple comparisons for the significant simple main effects of sweetener in 74 time windows.

Time (s)	Sucrose	Sucralose	Acesulfame K	Time (s)	Sucrose	Sucralose	Acesulfame K
3	2.78 [a]	2.81 [a]	2.56 [b]	77	0.89 [a]	1.23 [b]	1.06 [c]
5	3.43 [a]	3.44 [a]	3.16 [b]	79	0.86 [a]	1.19 [b]	1.02 [a]
7	3.44 [a]	3.45 [a]	3.16 [b]	81	0.82 [a]	1.15 [b]	0.99 [c]
9	3.28 [a]	3.34 [a]	3.03 [b]	83	0.78 [a]	1.11 [b]	0.97 [b]
11	3.10 [a]	3.21 [a]	2.88 [b]	85	0.75 [a]	1.09 [b]	0.95 [b]
13	2.97 [a]	3.05 [a]	2.76 [b]	87	0.72 [a]	1.06 [b]	0.93 [b]
15	2.85 [a]	2.92 [a]	2.67 [b]	89	0.69 [a]	1.02 [b]	0.90 [b]
17	2.71 [a,b]	2.81 [a]	2.56 [b]	91	0.67 [a]	0.99 [b]	0.89 [b]
19	2.60 [a,b]	2.71 [a]	2.46 [b]	93	0.64 [a]	0.95 [b]	0.85 [b]
21	2.51 [a,b]	2.62 [a]	2.36 [b]	95	0.62 [a]	0.90 [b]	0.81 [b]
23	2.41 [a,b]	2.54 [a]	2.30 [b]	97	0.59 [a]	0.87 [b]	0.79 [b]
25	2.30 [a]	2.46 [b]	2.23 [a]	99	0.56 [a]	0.84 [b]	0.76 [b]
27	2.20 [a]	2.41 [b]	2.16 [a]	101	0.53 [a]	0.82 [b]	0.74 [b]
29	2.11 [a]	2.35 [b]	2.11 [a]	103	0.53 [a]	0.79 [b]	0.71 [b]
31	2.04 [a]	2.31 [b]	2.04 [a]	105	0.51 [a]	0.76 [b]	0.68 [b]
33	1.97 [a]	2.25 [b]	1.97 [a]	107	0.49 [a]	0.75 [b]	0.66 [b]
35	1.90 [a]	2.18 [b]	1.91 [a]	109	0.48 [a]	0.73 [b]	0.63 [a,b]
37	1.86 [a]	2.12 [b]	1.84 [a]	111	0.47 [a]	0.71 [b]	0.60 [a,b]
39	1.77 [a]	2.07 [b]	1.80 [a]	113	0.45 [a]	0.69 [b]	0.58 [a,b]
41	1.70 [a]	2.03 [b]	1.73 [a]	115	0.43 [a]	0.67 [b]	0.56 [a,b]
43	1.63 [a]	2.00 [b]	1.67 [a]	117	0.42 [a]	0.66 [b]	0.54 [a,b]
45	1.59 [a]	1.94 [b]	1.61 [a]	119	0.41 [a]	0.65 [b]	0.53 [a,b]
47	1.55 [a]	1.89 [b]	1.57 [a]	121	0.39 [a]	0.63 [b]	0.50 [a,b]
49	1.49 [a]	1.84 [b]	1.54 [a]	123	0.36 [a]	0.61 [b]	0.48 [a,b]
51	1.45 [a]	1.80 [b]	1.51 [a]	125	0.34 [a]	0.59 [b]	0.47 [a,b]
53	1.42 [a]	1.74 [b]	1.46 [a]	127	0.32 [a]	0.57 [b]	0.46 [a,b]
55	1.37 [a]	1.68 [b]	1.43 [a]	129	0.30 [a]	0.56 [b]	0.45 [a,b]
57	1.33 [a]	1.64 [b]	1.40 [a]	131	0.29 [a]	0.55 [b]	0.44 [a,b]
59	1.27 [a]	1.60 [b]	1.36 [a]	133	0.28 [a]	0.52 [b]	0.42 [a,b]
61	1.23 [a]	1.56 [b]	1.33 [a]	135	0.28 [a]	0.49 [b]	0.41 [a,b]
63	1.17 [a]	1.51 [b]	1.27 [a]	137	0.27 [a]	0.48 [b]	0.41 [a,b]
65	1.12 [a]	1.46 [b]	1.23 [a]	139	0.26 [a]	0.47 [b]	0.41 [a,b]
67	1.07 [a]	1.42 [b]	1.20 [a]	141	0.26 [a]	0.46 [b]	0.40 [a,b]
69	1.03 [a]	1.38 [b]	1.17 [a]	143	0.25 [a]	0.45 [b]	0.40 [a,b]
71	0.99 [a]	1.33 [b]	1.15 [c]	145	0.24 [a]	0.44 [b]	0.39 [a,b]
73	0.95 [a]	1.29 [b]	1.12 [c]	147	0.24 [a]	0.43 [b]	0.38 [a,b]
75	0.92 [a]	1.26 [b]	1.09 [c]	149	0.23 [a]	0.42 [b]	0.37 [a,b]

Values for each time window are medians (e.g., 1 s means the time window from 0–2 s), and values for each sweetener are sweetness intensities. In each time window, values with same letters did not differ statistically, but values with different letters differed significantly with $p < 0.05$.

3.2. Comparison of Temporal Profiles between Solvent Sessions for Each Sweetener

3.2.1. Sucrose

Two-way repeated measures ANOVA revealed significant main effects of solvent session (F [1, 93] = 41.28, $p < 0.001$) and time (F [74, 6882] = 344.18, $p < 0.001$), and a significant interaction between solvent session and time (F [74, 6882] = 4.74, $p < 0.001$). The results of simple effects tests for interaction revealed significant simple main effects of solvent session in 70 time windows (medians of each time window = 11–149 s) and significant simple main effects of time in both solvent sessions ($p < 0.05$). These results indicated that in some time windows, the perceived intensity of sweetness was significantly higher for sucrose-sweetened coffee beverage than for sucrose-sweetened water solution.

3.2.2. Sucralose

Two-way repeated measures ANOVA revealed significant main effects of solvent session (F [1, 93] = 16.72, $p < 0.001$) and time (F [74, 6882] = 301.38, $p < 0.001$), and a significant interaction between solvent session and time (F [74, 6882] = 2.43, $p < 0.001$). The results of simple effects tests for interaction revealed significant simple main effects of solvent session in 64 time windows (medians of each time window = 9–129 s and 133–137 s) and significant simple main effects of time in both solvent sessions ($p < 0.05$). These results indicated that in some time windows, the perceived intensity of sweetness was significantly higher for sucralose-sweetened coffee beverage than for sucralose-sweetened water solution.

3.2.3. Acesulfame K

Two-way repeated measures ANOVA revealed significant main effects of solvent session (F [1, 93] = 36.17, $p < 0.001$) and time (F [74, 6882] = 300.36, $p < 0.001$), and a significant interaction between solvent session and time (F [74, 6882] = 4.48, $p < 0.001$). The results of simple effects tests for interaction revealed significant simple main effects of solvent session in 67 time windows (medians of each time window = 3–135 s) and significant simple main effects of time in both solvent sessions ($p < 0.05$). These results indicated that in some time windows, the perceived intensity of sweetness was significantly higher for acesulfame K–sweetened coffee beverage than for acesulfame K–sweetened water solution.

4. Discussion

4.1. Temporal Profiles of Sweetness of Sweetened Coffee Beverages

In this study, untrained panelists performed a time–intensity evaluation of sweetness using three coffee beverages and three water solutions. Temporal profiles of sweetness followed similar traces among all coffee beverages. To explain this result, we propose the following hypotheses. First, differences among sweeteners may have been masked by factors other than the sweetness of the sweeteners contained in the coffee beverages. This hypothesis would be supported by the observation of significant differences among sweeteners in the time–intensity evaluation of the sweetness of water solutions. Second, untrained panelists might be unable to perceive differences among sweeteners when performing a time–intensity evaluation of the sweetness of coffee beverages or water solutions. However, this second hypothesis was not supported because temporal profiles significantly differed among the water solutions, leaving the first hypothesis as a potential explanation. To further test the validity of the first hypothesis, for each sweetener we compared the temporal profiles of sweetness between a coffee beverage and a water solution. The results demonstrated that for all sweeteners, the temporal profile was significantly higher for the coffee beverage than for the water solution. In other words, factors other than the sweetness of a sweetener might affect the temporal profiles of sweetness of coffee beverages.

The following factors might affect the temporal profiles of sweetness of the coffee beverages. The first candidate is the sweetness of milk [37–39]. We hypothesized that the sweetness of the water solutions was derived only from sweetener, whereas the sweetness of the coffee beverages was derived from the addition of milk or the synergy between sweetener and milk. The second candidate is the aroma of milk [40–44]. We hypothesized that addition of milk caused a decrease in coffee-like aroma [45,46] and an increase in sweet aroma [47], thereby enhancing the sweetness. The third candidate is the "halo-damping effect" [48–50] of sensory attributes other than sweetness. When a participant performs psychophysical evaluation of a single sensory attribute per trial, the evaluation value might be biased by the effects of sensory attributes other than the one to which they were instructed to pay attention [13,14]. Therefore, we hypothesized that because participants performed a time–intensity evaluation of only sweetness in this study, the perceived intensity of sweetness increased due to the halo-damping effect. The fourth candidate is the novelty of the coffee beverages. We hypothesized that due to the use of unreleased products in this study, the novelty of the coffee beverages distracted the untrained panelists and prevent them from discriminating sweetness among sweeteners. The fifth candidate is the interactions between the sweetener and other ingredients. We hypothesized that such effects elicited a bitter taste. The sixth candidate is the viscosity of coffee beverages. Because coffee beverages are more viscous than water solutions, sweeteners contained in coffee beverages might bind more strongly to taste receptors, and might not be washed out as rapidly by the saliva. We hypothesized that this phenomenon might enhance lingering sweetness. The seventh candidate is the natural sweetness of coffee. Some reducing sugars are present in roasted coffee, so that the water-soluble part of them may contribute to the overall flavor of the coffee [51]. We hypothesized that the sweetness of coffee beverages was derived from the addition of a sweetener or the synergy between sweetener and coffee.

To model the temporal profiles of sweetness of beverages containing non-nutritive sweeteners on the temporal profiles of sweetness of sucrose-sweetened beverages, multiple non-nutritive sweeteners are used in combination [52]. Specially, sucralose and acesulfame K have been combined in many low-calorie foods and beverages to achieve a temporal profile of sweetness similar to that of sucrose [53–55]. Based on this trend, in the near future, we intend to employ a time–intensity evaluation of sweetness with untrained panelists using coffee beverages containing combinations of sucralose and acesulfame K.

4.2. Temporal Profiles of Sweetness of Sweeteners in Water Solutions

As controls in this study, we used three types of water solutions. The temporal profiles of sweetness significantly differed among these water solutions. This result was inconsistent with the results of some studies reporting that the duration of sweetness does not differ among sucrose, sucralose, and acesulfame K [17,18], and another study reporting that the duration of sweetness was longer for sucrose than for sucrose and acesulfame K [56]. The temporal profiles of sweetness of various sweeteners vary with concentration [57], and perceived intensities of sensory attributes differ among untrained and trained panelists [58,59]. These factors might explain why results differed between this and previous studies.

5. Conclusions

In this study, we investigated whether the temporal profiles of sweetness differed between an ordinary coffee beverage containing a nutritive sweetener and low-sugar coffee beverages containing non-nutritive sweeteners. A time–intensity evaluation of sweetness revealed that temporal profiles followed similar traces among sucrose-, sucralose-, and acesulfame K–sweetened coffee beverages. In other words, the result indicated that the sensory attributes of low-sugar products (coffee beverages containing non-nutritive sweeteners) are close to those of ordinary products (coffee beverage containing sucrose), at least from the viewpoint of lingering sweetness.

Additionally, to investigate why temporal profiles of sweetness did not differ among the three coffee beverages, the untrained panelists performed a time–intensity evaluation of sweetness using three water solutions (sucrose-sweetened, sucralose-sweetened, and acesulfame K–sweetened deionized water) as control samples. We observed significant differences among the temporal profiles of the three water solutions, suggesting that differences in the temporal profiles of the three sweetened coffee beverages were masked by factors other than the sweetness of the sweeteners. Because this study was performed using untrained panelists, who represent the end users of released beverages, the results may provide information useful for beverage development.

Acknowledgments: This study was mainly funded by Asahi Soft Drink Co., Ltd., which had no control over the interpretation, writing, or publication of this work. This study was also partially supported by JSPS KAKENHI Grant Numbers 26245073 and 16K04418.

Author Contributions: Naomi Gotow participated in the study design and coordination, performed the statistical analysis, and drafted the manuscript; Shinji Esumi and Hirofumi Kubota conceived of the study and participated in its design; Tatsu Kobayakawa, who is the corresponding author, conceived of the study, participated in its design and coordination, undertook most of the revisions, and supervised the drafting of the manuscript.

Conflicts of Interest: This study was mainly funded by Asahi Soft Drink Co., Ltd.; this sponsor had no control over the interpretation, writing, or publication of this work.

References

1. White, J.S. Sucrose, HFCS, and fructose: History, manufacture, composition, applications, and production. In *Fructose, High Fructose Corn Syrup, Sucrose and Health*; Rippe, J.M., Ed.; Humana Press: New York, NY, USA, 2014; pp. 13–33.
2. Clemens, R.A.; Jones, J.M.; Kern, M.; Lee, S.-Y.; Mayhew, E.J.; Slavin, J.L.; Zivanovic, S. Functionality of sugars in foods and health. *Compr. Rev. Food Sci. Food Saf.* **2016**, *15*, 433–470. [CrossRef]
3. Gardner, C.; Wylie-Rosett, J.; Gidding, S.S.; Steffen, L.M.; Johnson, R.K.; Reader, D.; Lichtenstein, A.H.; on Behalf of the American Heart Association Nutrition Committee of the Council on Nutrition, Physical Activity and Metabolism, Council on Arteriosclerosis, Thrombosis and Vascular Biology, Council on Cardiovascular Disease in the Young, & the American Diabetes Association. Nonnutritive sweeteners: Current use and health perspectives: A scientific statement from the American Heart Association and the American Diabetes Association. *Diabetes Care* **2012**, *35*, 1798–1808. [PubMed]
4. Roberts, M.W.; Wright, J.T. Nonnutritive, low caloric substitutes for food sugars: Clinical implications for addressing the incidence of dental caries and overweight/obesity. *Int. J. Denti.* **2012**, *2012*, 625701. [CrossRef] [PubMed]
5. Japan Soft Drink Association. Question and Answer about Soft Drinks: Please Tell me the Difference between "Low Sugar" and "Non-Sugar" Coffee Beverages. Available online: http://www.j sda.or.jp/ippan/qa_view. php?id=55&cat=1 (accessed on 5 December 2017).
6. Ministry of Health, Labor and Welfare. In *Nutrition Labelling Standards*. Available online: http://www.caa. go.jp/foods/pdf/syokuhin344.pdf (accessed on 5 December 2017).
7. European Union. Regulation (EC) No. 1924/2006 of the European Parliament and of the Council of 20 December 2006 on nutrition and health claims made on foods. *Off. J. Eur. Union* **2006**, *49*, L404/9.
8. Institute of Medicine (US) Committee on Examination of Front-of-Package Nutrition Rating Systems and Symbols. Appendix B FDA regulatory requirements for nutrient content claims. In *Front-of-Package Nutrition Rating Systems and Symbols*; Wartella, E.A., Lichtenstein, A.H., Caitlin, S., Boon, C.S., Eds.; National Academies Press: Washington, DC, USA, 2010.
9. Insel, P.M.; Turner, R.E.; Ross, D. Carbohydrates: Simple sugars and complex chains. In *Discovering Nutrition*, 3rd ed.; Jones and Bartlett Publishers: Sudbury, MA, USA, 2010; pp. 135–167.
10. Tharp, B.W.; Young, L.S. High-intensity sweeteners. In *Tharp & Young on Ice Cream: An Encyclopedic Guide to Ice Cream Science and Technology*; DEStech Publications: Pennsylvania, PA, USA, 2013; pp. 176–177.
11. DuBois, G.E. Saccharin and cyclamate. In *Sweeteners and Sugar Alternatives in Food Technology*; Mitchell, H., Ed.; Blackwell Publishing: Oxford, UK, 2006; pp. 103–129.
12. Di Monaco, R.; Miele, N.A.; Volpe, S.; Picone, D.; Cavella, S. Temporal sweetness profile of MNEI and comparison with commercial sweeteners. *J. Sens. Stud.* **2014**, *29*, 385–394. [CrossRef]

13. Meillon, S.; Urbano, C.; Schlich, P. Contribution of the temporal dominance of sensations (TDS) method to the sensory description of subtle differences in partially dealcoholized red wines. *Food Qual. Prefer.* **2009**, *20*, 490–499. [CrossRef]

14. Pineau, N.; Schlich, P.; Cordelle, S.; Mathonnière, C.; Issanchou, S.; Imbert, A.; Rogeaux, M.; Etiévant, P.; Köster, E. Temporal dominance of sensations: Construction of the TDS curves and comparison with time–intensity. *Food Qual. Prefer.* **2009**, *20*, 450–455. [CrossRef]

15. Lee, W.E., III; Pangborn, M. Time–intensity: The temporal aspects of sensory perception. *Food Technol.* **1986**, *40*, 71–78, 82.

16. Ayya, N.; Lawless, H.T. Potency of sweetness of aspartame, d-tryptophan and thaumatin evaluated by single value and time–intensity measurements. *Chem. Senses* **1992**, *17*, 245–259. [CrossRef]

17. Duizer, L.M.; Bloom, K.; Findlay, C.J. The effect of line orientation on the recording of time–intensity perception of sweetener solutions. *Food Qual. Prefer.* **1995**, *6*, 121–126. [CrossRef]

18. Ketelsen, S.M.; Keay, C.L.; Wiet, S.G. Time–intensity parameters of selected carbohydrate and high potency sweeteners. *J. Food Sci.* **1993**, *58*, 1418–1421. [CrossRef]

19. Ott, D.B.; Ledwards, C.; Palmer, S.J. Perceived taste intensity and duration of nutritive and non-nutritive sweeteners in water using time–intensity (T-I) evaluations. *J. Food Sci.* **1991**, *56*, 535–542. [CrossRef]

20. Galmarini, M.V.; Zamora, M.C.; Chirife, J. Gustatory reaction time and time intensity measurements of trehalose and sucrose solutions and their mixtures. *J. Sens. Stud.* **2009**, *24*, 166–181. [CrossRef]

21. Azevedo, B.M.; Schmidt, F.L.; Bolini, H.M.A. High-intensity sweeteners in espresso coffee: Ideal and equivalent sweetness and time–intensity analysis. *Int. J. Food Sci. Technol.* **2015**, *50*, 1374–1381. [CrossRef]

22. Rodrigues, J.B.; Paixão, J.A.; Cruz, A.G.; Bolini, H.M.A. Chocolate milk with chia oil: Ideal sweetness, sweeteners equivalence, and dynamic sensory evaluation using a time–intensity methodology. *J. Food Sci.* **2015**, *80*, S2944–S2949. [CrossRef] [PubMed]

23. De Souza, V.R.; Pereira, P.A.P.; Pinheiro, A.C.M.; Bolini, H.M.A.; Borges, S.V.; Queiroz, F. Analysis of various sweeteners in low-sugar mixed fruit jam: Equivalent sweetness, time–intensity analysis and acceptance test. *Int. J. Food Sci. Technol.* **2013**, *48*, 1541–1548. [CrossRef]

24. Palazzo, A.B.; Carvalho, M.A.R.; Efraim, P.; Bolini, H.M.A. The determination of isosweetness concentrations of sucralose, rebaudioside and neotame as sucrose substitutes in new diet chocolate formulations using the time–intensity analysis. *J. Sens. Stud.* **2011**, *26*, 291–297. [CrossRef]

25. Patil, S.; Ravi, R.; Saraswathi, G.; Prakash, M. Development of low calorie snack food based on intense sweeteners. *J. Food Sci. Technol.* **2014**, *51*, 4096–4101. [CrossRef] [PubMed]

26. De Melo, L.L.M.M.; Bolini, H.M.A.; Efraim, P. Equisweet milk chocolates with intense sweeteners using time–intensity method. *J. Food Qual.* **2007**, *30*, 1056–1067. [CrossRef]

27. Melo, L.; Bolini, H.M.A.; Efraim, P. Low-calorie chocolates and acceptability/sensory properties. In *Chocolate in Health and Nutrition*; Watson, R., Preedy, V.R., Zibadi, S., Eds.; Humana Press: New York, NY, USA, 2013; pp. 163–176.

28. Gotow, N.; Moritani, A.; Hayakawa, Y.; Akutagawa, A.; Hashimoto, H.; Kobayakawa, T. Development of a time–intensity evaluation system for consumers: Measuring bitterness and retronasal aroma of coffee beverages in 106 untrained panelists. *J. Food Sci.* **2015**, *80*, S1343–S1351. [CrossRef] [PubMed]

29. Japan Soft Drink Association. Question and Answer about Soft Drinks: Stevia. Available online: http://www.j-sda.or.jp/sp/qa_view.php?id=118&cat=8 (accessed on 5 December 2017).

30. Green, B.G.; Lim, J.; Osterhoff, F.; Blacher, K.; Nachtigal, D. Taste mixture interactions: Suppression, additivity, and the predominance of sweetness. *Physiol. Behav.* **2010**, *101*, 731–737. [CrossRef] [PubMed]

31. Lawless, H.T.; Schlake, S.; Smythe, J.; Lim, J.; Yang, H.; Chapman, K.; Bolton, B. Metallic taste and retronasal smell. *Chem. Senses* **2004**, *29*, 25–33. [CrossRef] [PubMed]

32. Gotow, N.; Moritani, A.; Hayakawa, Y.; Akutagawa, A.; Hashimoto, H.; Kobayakawa, T. High consumption increases sensitivity to after-flavor of canned coffee beverages. *Food Qual. Prefer.* **2015**, *44*, 162–171. [CrossRef]

33. Lawless, H.L.; Heymann, H. Context effects and biases in sensory judgement. In *Sensory Evaluation of Food: Principles and Practices*, 2nd ed.; Springer: New York, NY, USA, 2010; pp. 203–225.

34. Plemmons, L.E.; Resurreccion, A.V.A. A warm-up sample improves reliability of responses in descriptive analysis. *J. Sens. Stud.* **1998**, *13*, 359–376. [CrossRef]

35. Lim, J.; Johnson, M.B. Potential mechanisms of retronasal odor referral to the mouth. *Chem. Senses* **2011**, *36*, 283–289. [CrossRef] [PubMed]

36. Lim, J.; Johnson, M.B. The role of congruency in retronasal odor referral to the mouth. *Chem. Senses* **2012**, *37*, 515–522. [CrossRef] [PubMed]
37. Chapman, K.W.; Lawless, H.T.; Boor, K.J. Quantitative descriptive analysis and principal component analysis for sensory characterization of ultrapasteurized milk. *J. Dairy Sci.* **2001**, *84*, 12–20. [CrossRef]
38. Heymann, H.; Lawless, H.T. Context effects and biases in sensory judgment. In *Sensory Evaluation of Food: Principles and Practices*; Springer Science+Business Media: New York, NY, USA, 1999; pp. 301–340.
39. Lee, G.H.; Lee, J.S.; Shin, M.G. Sensory attribute comparison of consumer milk using descriptive analysis. *Food Sci. Biotechnol.* **2003**, *12*, 480–484.
40. Frank, R.A.; Byram, J. Taste–smell interactions are tastant and odorant dependent. *Chem. Senses* **1988**, *13*, 445–455. [CrossRef]
41. Frank, R.A.; Ducheny, K.; Mize, S.J.S. Strawberry odor, but not red color, enhances the sweetness of sucrose solutions. *Chem. Senses* **1989**, *14*, 371–377. [CrossRef]
42. Labbe, D.; Damevin, L.; Vaccher, C.; Morgenegg, C.; Martin, N. Modulation of perceived taste by olfaction in familiar and unfamiliar beverages. *Food Qual. Prefer.* **2006**, *17*, 582–589. [CrossRef]
43. Schifferstein, H.N.; Verlegh, P.W. The role of congruency and pleasantness in odor-induced taste enhancement. *Acta Psychol.* **1996**, *94*, 87–105. [CrossRef]
44. Stevenson, R.J.; Prescott, J.; Boakes, R.A. Confusing tastes and smells: How odours can influence the perception of sweet and sour tastes. *Chem. Senses* **1999**, *24*, 627–635. [CrossRef] [PubMed]
45. Bücking, M.; Steinhart, H. Headspace GC and sensory analysis characterization of the influence of different milk additives on the flavor release of coffee beverages. *J. Agric. Food Chem.* **2002**, *50*, 1529–1534. [CrossRef] [PubMed]
46. Itobe, T.; Nishimura, O.; Kumazawa, K. Influence of milk on aroma release and aroma perception during consumption of coffee beverages. *Food Sci. Technol. Res.* **2015**, *21*, 607–614. [CrossRef]
47. Liu, J.; Liu, M.; He, C.; Song, H.; Guo, J.; Wang, Y.; Yang, H.; Su, X. A comparative study of aroma-active compounds between dark and milk chocolate: Relationship to sensory perception. *J. Sci. Food Agric.* **2015**, *95*, 1362–1372. [CrossRef] [PubMed]
48. Clark, C.C.; Lawless, H.T. Limiting response alternatives in time–intensity scaling: An examination of the halo-dumping effect. *Chem. Senses* **1994**, *19*, 583–594. [CrossRef] [PubMed]
49. Prescott, J. Flavour as a psychological construct: Implications for perceiving and measuring the sensory qualities of foods. *Food Qual. Prefer.* **1999**, *10*, 349–356. [CrossRef]
50. Prescott, J.; Johnstone, V.; Francis, J. Odor-taste interactions: Effects of attentional strategies during exposure. *Chem. Senses* **2004**, *29*, 331–340. [CrossRef] [PubMed]
51. Heath, H.B. The Physiology of Flavour: Taste and Aroma Perception. In *Coffee: Volume 3, Physiology*; Clarke, R.J., Macrae, R., Eds.; Elsevier Applied Science Publisher: Essex, UK, 1988; pp. 141–170.
52. Bakal, A.I. Mixed Sweetener Functionality. In *Alternative Sweeteners*, 3rd ed.; Revised and Expand; O'Brien-Nabors, L., Ed.; Marcel Dekker, Inc.: New York, NY, USA, 2001; pp. 463–480.
53. Kemp, S.E.; Lindley, M.G. Developments in sweeteners for functional and speciality beverages. In *Functional and Speciality Beverage Technology*; Paquin, P., Ed.; Woodhead Publishing Limited: Cambridge, UK, 2009; pp. 39–54.
54. Kim, P. Sweetness sense. Sweeteners: Customizing Sweetness Profiles. In *Food Product Design*; Deis, R.C., Ed.; Virgo Publishing: Phoenix, AZ, USA, 2006; Volume 15, Number 11; p. 1.
55. Karstadt, M.L. Testing needed for acesulfame potassium, an artificial sweetener. *Environ. Health Perspect.* **2006**, *114*, A516. [CrossRef] [PubMed]
56. Goldsmith, L.A.; Merkel, C.M. Sucralose. In *Alternative Sweeteners*, 3rd ed.; Revised and Expand; O'Brien-Nabors, L., Ed.; Marcel Dekker, Inc.: New York, NY, USA, 2001; pp. 185–207.
57. Mahindru, S.N. High intensity-low calorie sweeteners. In *Food Additives: Characteristics, Detection and Estimation*; APH Publishing Corporation: New Delhi, India, 2008; pp. 53–104.

58. Losó, V.; Gere, A.; Györey, A.; Kókai, Z.; Sipos, L. Comparison of the performance of a trained and an untrained panel on sweetcorn varieties with the panelcheck software. *APSTRACT* **2012**, *1–2*, 77–83.
59. Roberts, A.K.; Vickers, Z.M. A comparison of trained and untrained judges' evaluation of sensory attribute intensities and liking of cheddar cheeses. *J. Sens. Stud.* **1994**, *9*, 1–20. [CrossRef]

Article

Dose-Response Relationships for Vanilla Flavor and Sucrose in Skim Milk: Evidence of Synergy

Gloria Wang [1,2], **John E. Hayes** [1,2], **Gregory R. Ziegler** [1], **Robert F. Roberts** [1] and **Helene Hopfer** [1,2,*]

[1] Department of Food Science, The Pennsylvania State University, University Park, PA 16802, USA; gzw19@psu.edu (G.W.); jeh40@psu.edu (J.E.H.); grz1@psu.edu (G.R.Z.); rfr3@psu.edu (R.F.R.)

[2] The Sensory Evaluation Center, The Pennsylvania State University, University Park, PA 16802, USA

* Correspondence: hopfer@psu.edu; Tel.: +1-814-863-5572

Received: 31 August 2018; Accepted: 26 September 2018; Published: 4 October 2018

Abstract: Regarding cross-modality research, taste-aroma interaction is one of the most studied areas of research. Some studies have reported enhancement of sweetness by aroma, although it is unclear as to whether these effects actually occur: depending on the cognitive strategy employed by panelists, the effects may disappear, e.g., forcing panelists into an analytical strategy to control for dumping may not be able to reveal perceptual interactions. Previous studies have largely focused on solutions and model foods, and did not test stimuli or concentrations relevant to real food applications. This study addresses these gaps: 18 vanilla flavored sucrose milks, varying between 0–0.75% (w/w) two-fold vanilla, and 0–5% (w/w) sucrose, were rated by 108 panelists for liking and perceived sweetness, vanilla flavor, milk flavor, and thickness. Interactions between vanilla and sucrose were measured using deviations of real mixtures from additive models (via the isobole method), indicating vanilla aroma does enhance perceived sweetness. However, the sweetness enhancing effect of vanilla aroma was not as pronounced as that of sucrose on vanilla flavor. Measurable cross-modal interactions occur despite using an analytical cognitive strategy. More work is needed to investigate the influence of perceptual strategy on the degree of taste-aroma interactions in real foods.

Keywords: cross-modality; taste-aroma interactions; sweetness enhancement; vanilla flavor; flavored milk; sugar; isoboles; synergy

---

1. Introduction

Flavor perception is the result of chemical and physical food properties, and how they interact with our senses [1]. Flavor involves the integration of multiple modalities, including smell, taste, and touch, and studying each in isolation does not reflect what humans experience during eating. As a result, cross-modal interactions (i.e., the interaction of taste, aroma, vision, texture, and chemesthesis) are a popular area of study in flavor research. Taste-aroma interactions are the most commonly described interaction between sensory modalities, and occur as a result of physical, physiological, cognitive, and psychological effects [2,3]. Research on multisensory processes, including taste-aroma interactions, has been used to better explain the processes humans use to assess food flavor [4–6]. Due to the common confusion between smell and taste, taste perception is influenced by odor and vice versa [7–11]. In order for perceived taste intensity to be modified by an odor (or vice versa), not only are the method of stimulation and the instructions given important, but also the perceptual similarity between a tastant and odorant [3,10]. As defined by Schifferstein and Verlegh [12], the extent to which two stimuli interact in combination in a food is called congruency. Some work on taste-aroma interactions suggested that sweetness intensity can be enhanced by a congruent odor, although many of these studies have only been done in model sucrose solutions [7,9–11,13–21] or model foods [22–26].

Due to the complexity of food products, only a few taste-aroma interaction studies have been conducted in real foods, such as whipped cream, milk, fruit juices, ciders, custards and cherry drinks [7,27–30]. Although these studies moved taste-aroma interaction research into a more realistic matrix, the concentration ranges for both the odor and the taste component used in these studies was not always commercially relevant. This may be important since odor compounds have been shown to impact flavor perception at both subthreshold and suprathreshold concentrations, i.e., taste-aroma interactions have also been shown to be concentration dependent [29,31–33]. The lack of studies in real food products over a concentration range comparable to those used in commercial products prevents a comprehensive understanding and utilization of such cross-modal interaction phenomena in real foods.

Fluid milk is a relatively simple food well suited for studying taste-aroma interactions. While plain milk consumption has been on the decline, flavored milk consumption has been increasing, and is expected to continue to grow as flavor becomes more important to consumers, notably in the US [34]. A familiar congruent aroma-taste pair in Western cultures used in numerous dairy products is vanilla and sucrose. Vanilla is the most popular flavor for dairy applications such as yogurt and ice cream, and is a complementary ingredient in flavored milks [35]. The combination of vanilla and sucrose is also commonly studied in cross-modal research [36–38]; therefore, using vanilla and sucrose in a flavored milk application increases the ecological validity of studying cross-modal interactions in foods.

Despite an abundance of literature suggesting an enhancement of perceived sweetness by a congruent aroma (including vanilla), there are other contradictory studies as well. This discrepancy exists because assessment of taste-aroma interactions strongly depends on the cognitive task (i.e., test questions and instructions) used by assessors when evaluating samples. Previous studies using rating scales have shown reduced or no mixture-induced taste enhancement when assessors are asked to evaluate perceived sweetness as well as perceived aroma intensity [7,10]. Thus, some researchers have attributed any sweetness enhancement to a "dumping effect" [37]. That is, when a scale for a pertinent flavor attribute is not provided in the test (e.g., vanilla flavor), assessors will "dump" their perceptions into another similar category (e.g., sweetness) instead, leading to an increase in attribute intensity [37]. In contrast to this analytical mindset, adopting a synthetic mindset is more in-line with real eating behavior where the food is experienced as a whole and a hedonic and holistic evaluation of the food occurs. In the analytical mindset, enhancement effects have been found [7,13,15,17,18,27,39], but in many of these early studies participants did not rate the intensity of all salient product attributes (taste, aroma, and texture), which may have resulted in dumping. This makes them somewhat inconclusive, raising the question of whether enhancements or any interactions truly occur.

To address these unresolved questions, additional experiments that provide a full range of relevant product attributes are needed. Several studies have found strong evidence for odor (but not taste) enhancements in aqueous solutions and food matrices, even when controlling for dumping. For example, Green and colleagues [30] found enhanced odor perception in solutions, vanilla custard, and a cherry drink, by the addition of sucrose. Lim's group [40] found a similar effect for citral and sucrose solutions, and later [41] for citral and coffee solutions. Although these studies demonstrated that an analytical strategy did not prevent taste-aroma interactions, their use of a sip and spit procedure is not representative of normal eating behavior, which has been shown to influence the sensory profile of samples, depending on the taste and flavor characteristics [42,43].

Separately, testing for interaction between mixture components also requires a different model approach above and beyond simple significance testing of mean attribute ratings. Instead, one needs to test whether the degree of interaction between two stimuli is above (or below) what would be predicted in an additive model. A common approach for testing drug interactions is the isobole method [44–48]. The isobole approach uses concentrations, effects, and empirical concentration-effect relationships, and is independent of the mechanism of interaction; instead, it is based on the concept of concentration addition. The points on an isobole indicate the mixing values of multiple components at which a specific quantitative effect is produced that is either synergistic, antagonistic, or simply additive.

Isoboles have been used in a few studies to measure the interaction between food ingredients on chemosensation [49–51]. As the isobole approach is a useful concept for evaluating the type and degree of interactions between substances, regardless of their mechanisms of action [48], cross-modality of a sweet tastant and a congruent aroma can thus also be described and tested with the isobole method.

This study aimed to address the gaps outlined above by: (i) conducting a dose-response experiment in fluid milk to measure the cross-modal interactions between a sweet tastant and a congruent aroma; (ii) using a complete concentration design space relevant for flavored milks; (iii) controlling for potential dumping effects; and (iv) using the isobole method to measure the type and degree of interaction.

2. Materials and Methods

2.1. Experimental Design and Sample Production

Eighteen sucrose-vanilla combinations were generated to model human sensory responses to the various mixtures with higher-order models (Table A1). Part of the design was composed of four levels across each of the two ingredients (in half-log steps) to generate dose-response functions for sucrose (granulated cane sugar) in milk and vanilla in milk. Additionally, a 3 × 3 factorial design was overlaid to create sucrose-vanilla mixtures in milk to generate the response surfaces for each sensory attribute. Collectively, this provides a total of 17 milk samples at systematically varied combinations of vanilla and sucrose. An additional control (plain milk) was added to bring the total number to 18. Milks were formulated with sucrose ranging 0–5% (w/w) and two-fold vanilla extract (i.e., twice the quantity of vanilla beans extracted in water and ethanol) ranging 0–0.75% (w/w), spanning a wide range of sucrose and vanilla concentrations, including those used industrially [52]. Two-fold vanilla extract as opposed to single-fold was used to minimize the amount of extractives added to the samples.

All milks were mixed with varying levels of sucrose in eight 32 kg-batches for pasteurization. Following pasteurization, 10 kg batches of each sucrose-vanilla combination were produced for sensory testing. Two-fold vanilla extract (David Michael & Co, now Tastepoint by International Flavors and Fragrances; Philadelphia, PA, USA) was used as the vanilla flavor. Pasteurized skim milk (0.18% fat, 8.91% solids) and sucrose (Golden Barrel, Honey Brook, PA, USA) were provided by the Berkey Creamery (University Park, PA, USA). Prior to pasteurization, all amounts of milk and sugar were pre-weighed into stainless steel milk cans and plastic tubs, respectively. On the day of pasteurization, sugar was dissolved into one third of the milk from each milk can and then transferred back to the milk can for further mixing with metal agitators. Sucrose-milk premixes were then blended into each milk can by mixing with metal agitators on high speed for 10 min for complete dispersion and dissolution of the sugar. All mixes were pasteurized (high temperature short time (HTST), APV Junior Pasteurizer, APV Invensys, Woodstock, GA, USA) at 75 °C for 25 s, and homogenized (Gaulin, Lake Mills, WI, USA) in a two-stage process at 10.3 and 3.5 mPa (2000 and 500 psi), cooled (<7 °C), and collected into milk cans. From there, 13 mixes were flavored with pre-weighed vanilla extract and divided into 17 different sucrose-vanilla combinations. Each milk was packaged into $\frac{1}{2}$-gallon opaque plastic milk jugs and stored at refrigeration temperature (<5 °C) for 5–7 days prior to sensory testing. All 18 samples were collected for physical analysis (percent total solids and percent fat; SMART Trac, CEM Corporation, Matthews, NC, USA) to ensure sucrose concentrations were within the required ranges. All 18 samples were also tested for coliforms (high-sensitivity Petrifilm; 3M, Maplewood, MN, USA) to ensure samples were suitable for human consumption. Viscosity of all milk samples was tested to account for potential physicochemical interactions that may lead to differences in flavor perception.

The viscosity was calculated as the slope of the shear stress vs. shear strain rate flow curve with shear strain rate ranging from 0 to 100 s^{-1} at 5 °C to mimic sample serving temperature. The flow curves were plotted with Trios Software (TA Instruments, New Castle, DE, USA), omitting stress overshoot/noise at low shear strain rate (0 to 15 s^{-1}). Two measurements for each sample were taken on a Discovery H3 Hybrid rheometer (TA Instruments, New Castle, DE, USA), equipped with a double wall concentric cylinder (inner diameter 40.77 mm, outer diameter 43.88 mm, cup diameter 30.21 mm; Table A1).

2.2. Consumer Acceptability and Intensity Ratings

A central location test was conducted with 108 participants (women = 76, age = 19–71) over three days; panelists tasted six milk samples per day in complete block design. Sample presentation was counterbalanced across panelists using a modified Williams-Latin Square design, presenting each sample at each position 6–7 times, and all 18 samples were served on each day to avoid potential confounding between samples and test days. Approximately 45–50 mL of milk were poured into 3.25-oz. (~96 mL) plastic cups (Fabri-Kal, Kalamazoo, MI, USA) and lidded; cups were labeled with random three-digit blinding codes. All data were collected with Compusense Cloud (Compusense Inc., Guelph, ON, Canada).

Participants were screened for dietary restrictions, food allergies and product use (i.e., those who consumed skim or 1% milk at least once a week, and indicated they would be interested in tasting vanilla flavored milk). Procedures were exempted from institutional review board review by professional staff in the Penn State Office of Research Protections under the wholesome foods exemption in 45 CFR 46.101(b) (protocol number 33164). Participants provided informed consent prior to testing and were compensated for their time. All samples were served at 5 °C and tasted in individual tasting booths under red light and at ambient temperature (~21–22 °C). Panelists were given deionized water at room temperature (20 °C) to rinse before and in between each sample. Degree of liking for each sample was measured first on a nine-point hedonic scale, with labels for each value ranging from "like extremely" to "dislike extremely" [53]. Panelists then rated the perceived intensities for sweetness, vanilla flavor, milk flavor, and thickness on an unstructured line scale (valued from 0 to 100), anchored with "very weak" on the left and "very strong" on the right.

2.3. Statistical Analysis

Data were analyzed using R (version 3.3.3) [54] and RStudio (version 1.1.453, Boston, MA, USA). A two-way Analysis of Variance (ANOVA) ($p < 0.05$) with all two-way interactions for sucrose and vanilla as fixed effects on viscosity and all sensory attributes was conducted, and followed by calculation of Tukey's honestly significant differences (HSD) with the agricolae package (version 1.2-8) [55]. The rsm package (version 2.8) [56] was used to create 3D response-surface plots of percent sucrose, percent vanilla, and each sensory attribute, modelling up to a second-order regression model by multiple linear regression, and using unaveraged responses from each assessor. Final equations were chosen based on lack of fit testing ($p > 0.05$), variance explained and overall fit [57]. Using the isobole approach [58], the degree of interaction between sucrose and vanilla was calculated for perceived sweetness ratings. According to the criteria set by Suhnel, $I = 1$ indicates no interaction, $I < 1$ indicates synergism, and $I > 1$ indicates antagonism. In the equation, "c_i" denotes the concentration of component "i" in the mixture, and "C_i" represents the concentration of "i" that would individually produce the same intensity as the mixture, calculated from the corresponding dose-response functions of vanilla-only and sucrose-only milks as found by multiple linear regression of the non-mixture samples. Since the milk samples consist of sucrose and vanilla, the general equation can be rewritten as:

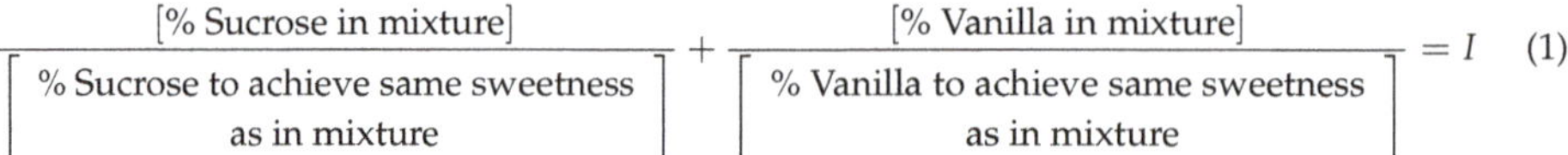

$$\frac{[\% \text{ Sucrose in mixture}]}{\left[\begin{array}{c}\% \text{ Sucrose to achieve same sweetness} \\ \text{as in mixture}\end{array}\right]} + \frac{[\% \text{ Vanilla in mixture}]}{\left[\begin{array}{c}\% \text{ Vanilla to achieve same sweetness} \\ \text{as in mixture}\end{array}\right]} = I \quad (1)$$

Three-dimensional isobolograms were generated for perceived sweetness using OriginPro (Origin Lab Corporation, version 2017 64-bit 94E, Northampton, MA, USA). To generate these plots, linear regression on the sweetness ratings found for the vanilla-only and sucrose-only samples was applied to generate a 3D sweetness plane. The mean attribute ratings for each of the sucrose-vanilla mixtures were then plotted onto the plane. Any points above the plane indicate synergism, while any values that fall below the plane are indicative of antagonism, and any that contact the plane indicate no interaction. Variation within the *I* values was accounted for by establishing a range of $0.9 < I < 1.1$ ($=\pm10\%$) as the zero-interaction criterion, similar to Fleming et al. [49].

3. Results

3.1. Physical Characterization

Overall, milk samples differed significantly in instrumental viscosity ($p < 0.05$), but only between the highest (5%) and lowest (0%) sucrose concentrations (Table A1). The observed viscosity differences were minor compared to the other factors, and were reflected in the sensory measurement of thickness, which also only differed significantly between the same high and low sucrose samples (Table A1).

3.2. Response Surface Models for Vanilla-Sucrose Milks

Three-dimensional response surface models were created for overall liking and perceived sweetness, vanilla flavor, milk flavor, and thickness (Figure 1). In general, the experimental design space used here was effective in quantifying potential interactions between sucrose and vanilla and their effects on sweet taste and liking in milk across a wide range of concentrations. Sweet taste, milk flavor, and thickness were all sufficiently modeled by a first-order model, indicating a linear relationship, while for overall liking and perceived vanilla flavor intensity a second-order model best represented the observed responses. Second-order models for liking and vanilla flavor perception were expected as both often follow an inverted U-shaped optimum: at too high sugar concentrations a product becomes too sweet for consumers. Likewise, vanilla dosing can have a strong effect on its flavor profile. Prior reports suggest that at low levels, vanilla extract adds slight modification notes, regular dosages impart the characteristic and expected flavor profile, while at high levels, off-notes (e.g., woody, phenolic, and alcoholic notes) may negatively influence liking [59].

For the three linear models for sweetness, milk flavor, and thickness, an increase in sucrose was accompanied by a significant increase in sweetness ($R^2 = 0.37$, $F(2, 1941) = 567.5$, $p < 0.05$; m = 10.54 for sucrose, m = 5.11 for vanilla) and thickness ($R^2 = 0.022$, $F(2, 1941) = 23.2$, $p < 0.05$; m = 1.97 for sucrose, m = -2.42 for vanilla), and a decrease in milk flavor ($R^2 = 0.015$, $F(2, 1941) = 16.24$; m = -1.04 for sucrose, m = -10.37 for vanilla) (Figure 1b,d,e). These findings were all expected and are similar to those found for coffee-flavored milks [60], where both thickness and milk flavor were influenced by sucrose and coffee extract concentration, respectively. That is, increasing sugar concentration led to higher viscosities, and both vanilla and coffee flavor extracts decreased/masked milk flavor.

For the two second-order models, both liking and vanilla flavor increased up to a concentration after which ratings for both decreased. For liking, the design space covered the preferred concentration range, as the optimal liking, i.e., the maximal point on the liking surface, was found at 3.79% sucrose and 0.53% vanilla (Figure 1a). When compared to the maximum liked sample served to participants, this optimum was quite close in terms of sucrose concentration (3.82%), but was slightly lower than the tasted sample's vanilla concentration of 0.625%. The maximal point for the second-order model for vanilla flavor was found to be at 4.98% sucrose and 0.52% vanilla (Figure 1c), which is above the highest tested sucrose-vanilla mixture in terms of sucrose concentration (3.82%) and just below the highest tested sucrose concentration of 5%, but within the range of tested vanilla concentrations of 0.3%, 0.435% and 0.625%.

Inspecting the response surfaces, it becomes apparent from the first-order model for sweetness that increasing the vanilla concentration leads to a slight but statistically significant enhancement of sweetness (coefficient = 5.11; $p < 0.05$). This can be seen in Figure 1b as the bottom and top edges of the plane are tilted slightly upward as vanilla concentration increases). Conversely, the addition of sucrose increased the perception of vanilla significantly and substantially in the regression model for vanilla flavor (coefficient = 11.25; $p < 0.05$) (Figure 1c). These findings are in agreement with Green et al. [30] and Welge-Lussen et al. [61] who found that the perception of retronasal vanilla odor was driven to a larger degree by the addition of sucrose than the addition of more vanilla flavor. Similarly, Alcaire et al. [62] found a decrease in vanilla flavor perception when a vanilla flavored milk dessert was reduced in sugar by 20%. The same pattern has been observed for other tastes: Linscott and Lim [63] found that the tastes impacted odor perception more than the odors affected tastes when pairing NaCl and monosodium glutamate (MSG) with chicken and soy sauce odors, respectively. Earlier, Pfeiffer et al. [14] reported synergistic effects of sucrose and acid on strawberry flavor, and that strawberry flavor was perceived more intensely when sucrose and acid were used together. Furthermore, it seems unlikely that these findings could be explained by a physiochemical "salting out" effect from the addition of sugar, as three different studies found that instrumentally measured volatile concentration and release profiles remained constant even when tastants concentrations (e.g., sucrose or acids) were varied [14,23,64]. Generally, our findings agree with others who reported significant odor enhancement by taste [40,41,65], and seem to indicate a cognitive basis for the observed flavor enhancement rather than physicochemical interactions that influence to the solubility of components in the vanilla milk mixtures.

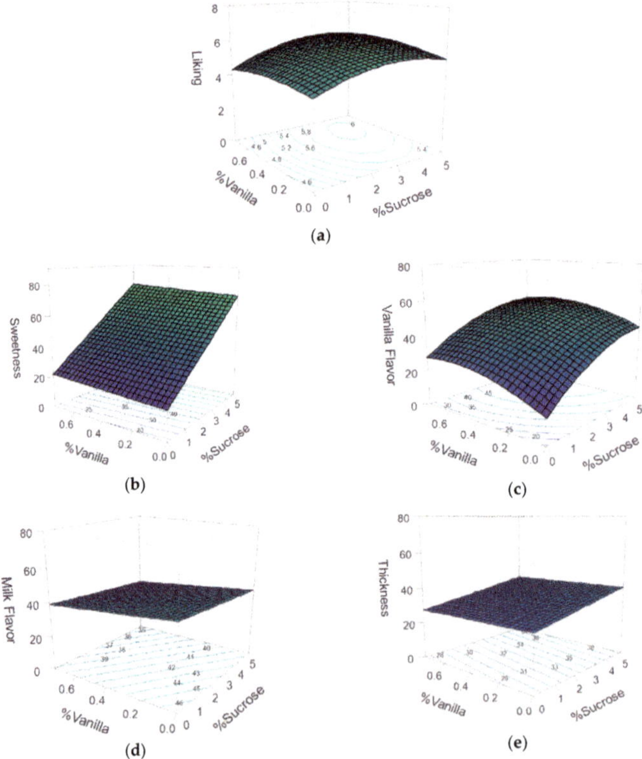

Figure 1. 3D response surfaces established from the 18 samples evaluated by 108 consumers, for all sensory response variables and sucrose (S) and vanilla (V) concentration. Dose-response functions are included for each surface: (**a**) overall liking OL = 4.40 + 0.669·S + 1.42·V + 0.198·S·V − 0.102·S^2 − 2.05·V^2; R^2 = 0.11, F(5, 1938) = 47.92, $p < 0.05$; (**b**) perceived sweetness SW = 18.1 + 10.5·S + 5.11·V; R^2 = 0.37, F(2, 1941) = 567.5, $p < 0.05$; (**c**) perceived vanilla flavor VF = 13.8 + 11.3·S + 50.6·V − 1.17·S − 1.07·S^2 − 42.6·V^2; R^2 = 0.21, F(5, 1938) = 104.7, $p < 0.05$; (**d**) perceived milk flavor MF = 46.9 − 1.04·S − 10.4·V; R^2 = 0.02, F(2, 1941) = 16.24, $p < 0.05$; and (**e**) perceived thickness TH = 29.4 + 1.97·S − 2.42·V; R^2 = 0.02, F(2, 1941) = 23.2, $p < 0.05$.

3.3. Testing for Interactions with the Isobole Approach

The response surface models indicated deviations from simple additive effects when sucrose and vanilla are combined; however, the degree of such interactions need to be formally tested with a method that accounts for the shape of the zero-interaction response surface (i.e., the isobole method). In Table 1, the indices of interaction (I) for sweetness, as calculated by Equation (1), are shown for all nine sucrose-vanilla mixtures. If using Suhnel's [58] liberal criterion for interaction (i.e., any $I \neq 1$), every mixture except for two (S3.82V0.3 and S3.82V0.625) exhibited synergism for perceived sweetness ($I < 1$), with the other two showing antagonism ($I > 1$) and no interaction ($I \sim 1$). However, applying our more realistic cut-off criterion, four samples at the low to medium sugar concentrations (1.31% and 2.24% sucrose at both 0.3% and 0.435% vanilla) showed evidence of synergy, with I values between 0.776 and 0.886. Further, interactions between sucrose and vanilla were stronger at lower vanilla concentrations, as observed for both sucrose concentrations. This indicates that lower levels of vanilla induce larger cross-modal interactions with sugar.

Table 1. Indices of interaction (I) for perceived sweetness in the nine sucrose-vanilla mixtures. S = sugar, V = vanilla, Cs and Cv represent the concentration of sugar and vanilla, respectively, that would individually produce the same intensity as the vanilla-milk mixture. C_S and C_V were calculated using the dose-response functions for sweetness with the sucrose-only mixtures (Sweetness $SW_{sugar} = 17.35 + 10.52 \cdot S$; $R^2 = 0.43$, F(1, 538) = 406.24, $p < 0.05$) and the vanilla-only mixtures (Sweetness $SW_{vanilla} = 15.71 + 6.41 \cdot V$; $R^2 = 0.007$, F(1, 538) = 4.92, $p < 0.05$); I was calculated using Equation (1).

Sample	Sucrose (% *w/w*)	Vanilla (% *w/w*)	Sweetness (-)	Cs	Cv	I
S1.31V0.300	1.31	0.300	37.4	1.91	3.38	0.776 [+]
S1.31V0.435	1.31	0.435	37.2	1.89	3.35	0.823 [+]
S1.31V0.625	1.31	0.625	36.3	1.80	3.21	0.922 [0]
S2.24V0.300	2.24	0.300	46.8	2.80	4.85	0.862 [+]
S2.24V0.435	2.24	0.435	46.9	2.81	4.87	0.886 [+]
S2.24V0.625	2.24	0.625	44.8	2.61	4.54	0.997 [0]
S3.82V0.300	3.82	0.300	58.8	3.94	6.73	1.01 [0]
S3.82V0.435	3.82	0.435	60.9	4.14	7.06	0.984 [0]
S3.82V0.625	3.82	0.625	58.1	3.88	6.62	1.08 [-]

[+] synergistic interaction; [0] zero interaction, [-] antagonistic interaction.

A more intuitive, graphical representation of these interactions is provided in Figure 2. Here, a zero-interaction response surface (the two-dimensional plane depicted in blue) was produced from the individual dose-response functions of the vanilla-only (Sweetness $SW_{vanilla} = 15.71 + 6.41 \cdot V$) and sucrose-only (Sweetness $SW_{sugar} = 17.35 + 10.52 \cdot S$) samples. Observed responses for specific vanilla-sucrose mixtures are visualized by individual dots; any mixtures that are located above the plane indicate synergism ($I < 1$), any mixtures below the plane are indicative of antagonism ($I > 1$), and any that lay on the zero-interaction surface demonstrate no interaction. Similar to the I-values shown in Table 1, all but two mixtures are above the additive plane, in line with Suhnel's criteria, and four of these also satisfied our working criterion (mixtures circled in black).

Overall, synergism seems to occur at the lower to medium sucrose and vanilla concentration ranges as suggested by the lower interaction values, indicating a concentration dependence effect of the vanilla and sugar on sweetness perception.

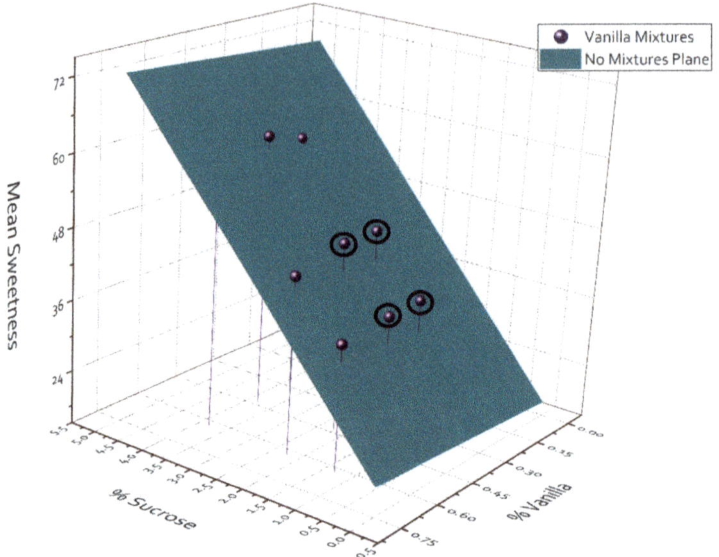

Figure 2. Isobologram for the vanilla-sucrose milks determined from their respective dose-response functions. Deviations above the zero-interaction surface correspond to *I*-values below 1 (synergy). Deviations below the surface correspond to *I*-values above 1 (antagonism). Those on or near the surface indicate no interaction (*I* = 1). Mixtures that satisfied the working criterion are circled in black.

4. Discussion

Overall, this study demonstrates in a real food product how closely sweet taste is associated with vanilla, i.e., the combination of vanilla and sugar creates a stronger vanilla flavor than vanilla alone. Vanilla also increases the perceived sweet taste of sugar, but the effect is smaller. It has consistently been reported that vanilla odor smells "sweet", at least by western participants, even though the olfactory system does not contain any receptors for sweet tastants [16,66]. This induction of sweetness by odors, even when they cannot be detected by taste receptors, seems related to previous combined exposure to vanilla aroma and a sweet tastant, such as those that could occur during eating [16,20,67].

Even when panelists were given a presumably complete list of attributes to rate (in an attempt to account for potential dumping effects), we still found evidence for interaction between the taste and the aroma. That is, adopting an analytical perceptual strategy did not appear to prevent taste enhancements from occurring. Aroma had a significant impact on taste and vice versa, though taste induced-odor enhancement appeared larger than odor induced-taste enhancement. Our findings are similar to other studies in that increasing sucrose concentration resulted in higher ratings of flavor [32,33,36,68]. Further, it should be noted that although increasing vanilla led to a significant sweetness enhancement, the magnitude of this taste enhancement by a congruent odor may be small in comparison to the use of a non-nutritive sweetener.

Despite results from other literature suggesting that any sweetness enhancement by an odor may be the result of a dumping artifact, our findings do not support that view. Rather, we were able to measure a small, but significant effect of vanilla aroma on sweetness perception. Valentin and Nguyen [69], when studying the effect of vanilla and lemon aroma on sourness ratings using two different tasks, came to a similar conclusion: providing more appropriate scales decreases odor-induced taste enhancement, but does not completely suppress it. Tasks that do not depend on intensity ratings have similarly found evidence of cross-modal interactions: in a sweet taste detection task, sucrose detection was increased when combined with a congruent strawberry odor versus an incongruent ham odor [70]. Similarly, in a categorization task, participants' categorization performance was poorer when the concentration of vanillin changed along with the sucrose concentration, implying that participants were unable to ignore the olfactory component when trying to categorize solutions according to the tastants [71]. Collectively, these studies and present results indicate that taste-odor interactions cannot be explained solely by response biases [72].

We also provide novel evidence of synergistic effects of mixture components by testing for interactions effects that go beyond the simple additivity by using the isobole method. Of the vanilla-sucrose milk mixtures we tested, four had indices of interaction (*I*-values) of less than 0.9 (a conservative cutoff), indicating a synergism when the two components were mixed together. As the isobole method was adapted from pharmacology, it would be important to determine where the index of interaction (*I*) provides a meaningful sensory effect. Future experiments with mixtures of varying *I* values are needed to determine differences that are perceivable by consumers. As Suhnel's criterion is calculated based on empirical data collected separately for each food matrix, it would be important to test whether the cut-off criteria used here generalizes across products. Based on our data, taste-aroma interactions are more likely to occur at low to medium sucrose-vanilla concentrations, which is in line with others (e.g., [73]) who similarly have found a concentration dependence.

Since taste and smell involve distinct mechanisms and do not share receptors, any interactions that take place would seem to be cognitive and not peripheral [3]. Neuroimaging studies have supported this view, in that certain brain regions are subsequently activated by taste-aroma combinations. For example, presenting a taste-odor combination activates the anterior region of the orbitofrontal cortex (OFC), a region that is not activated by taste or smell alone [38]. Similarly, blood oxygen level demand showed superadditivity in the OFC and amygdala for taste-odor combinations, compared to the individual components [74], and superadditivity of neural activity was also found in the OFC, insula, and anterior cingulate cortex (ACC) regions when presented with bimodal stimulation [38]. Additionally, congruency between tastants and odor further increases activation of these brain areas, suggesting that prior exposure is a necessary prerequisite for taste-aroma interactions [75]. These results all indicate that the combination of taste and smell leads to effects that are larger than the sum of their perceptual parts [1]. Even when the mathematical determination of synergism might be unclear, synergistic effects of tastants and odors are supported by neuroimaging research.

While the enhancement of taste by a congruent odor in this study was not strong, this may also be attributed to the cognitive task used. Early evidence of the dependence on cognition was found in 1984, when Lawless and Schlegel came to very different conclusions, depending on whether their panelists were given a direct or indirect scaling task to evaluate taste-odor mixtures. It was thought that panelists treat flavor mixtures as integral sensations during discrimination tasks, rather than as an analyzable pair of attributes during attribute rating [76]. This implies that the testing method induces different perceptual strategies: the discrimination task may reveal interactions between flavor notes, which would otherwise not be apparent when ratings are made of individual characteristics [11]. McBurney [77] stated that whether a taste-odor mixture is perceived analytically or synthetically can be determined by the method of evaluation required from the subject. The resulting elimination of or at least reduction in taste enhancement by the aroma occurs when one is forced to adopt an analytical approach once ratings of multiple attributes of a sample are required (thereby supporting the dumping phenomenon), whereas an overall evaluation of a food product encourages the synthesis of the common quality from both sensory modalities of odor and taste, causing enhancement [20].

Because the occurrence of an enhancement may result from scaling artifacts, a non-scaling method may be better suited for confirming the enhancement effects observed here. Using a more holistic (synthetic) testing method, consumers may perform more similarly to their "natural" behavior and, in doing so, reveal a greater taste enhancement. Such tests should be further explored, as the cognitive task during rating of various attributes does not represent how we perceive flavor when we eat food outside of such a controlled setting. To our knowledge, only a single study [77] has employed such a synthetic tactic to test for sweetness enhancement by using a matching task to control for possible scaling biases. More non-scaling studies are needed in cross-modal interaction research, as flavor is experienced as a singular percept rather than a series of individual components.

5. Conclusions

Even when controlling for potential dumping artifacts by providing all salient response categories to participants, significant interactions between vanilla and sugar were found for sweetness perception in a real food (skim milk). Use of a superadditive model, the isobole approach, provided further evidence for synergism in sweetness perception between a congruent taste-aroma pair, although these effects appear to be rather small in magnitude. Potential applications for sugar reduction in foods could be accomplished with these findings, although it probably requires combination with other current techniques, such as the use of non-nutritive sweeteners. Future work should test the most synergistic combinations with a holistic, non-scaling method, to better represent how we experience flavor when eating.

Author Contributions: Conceptualization, G.W., J.E.H., G.R.Z., R.F.R. and H.H.; Methodology and Software, G.W., J.E.H. and H.H.; Formal Analysis, G.W. and H.H.; Investigation, G.W.; Resources, H.H.; Writing—Original Draft Preparation, G.W.; Writing—Review and Editing, J.E.H., G.R.Z., R.F.R. and H.H.; Supervision, G.R.Z., R.F.R. and H.H.; Project Administration, H.H.; Funding Acquisition, G.W. and H.H.

Funding: This work was supported by the College of Agricultural Sciences at The Pennsylvania State University and USDA National Institute of Food and Agriculture Federal Appropriations under Project PEN04624 and Accession number 1013412. The funders had no role in the design of the study; in collection, analyses, or interpretation of data; in the writing of the manuscript, and in the decision to publish the results.

Acknowledgments: We thank the staff and students of the SEC and the Department of Food Science for assistance with sample processing, preparation and sensory experiments, with a special thank you to Tiffany Murray, Alyssa Bakke, Emily Furumoto, and Federico Harte. We are also thankful to Tastepoint by International Flavors and Fragrances (formerly David Michael & Co.) for donating the vanilla extract.

Conflicts of Interest: The authors declare no conflict of interest.

Appendix A

Table A1. Mean sensory attributes (n = 108) and measured instrumental viscosities (n = 2) (Pa.s) of final vanilla sucrose milks used in the dose-response study. S = sugar, V = vanilla, both in percent (w/w).

Sample	S	V	Overall Liking	Sweet Taste	Vanilla Flavor	Milk Flavor	Thickness	Viscosity (Pa.s)
S0V0	0.000	0.000	4.49 [fg]	14.8 [g]	12.9 [i]	50.5 [a]	30.3 [bcde]	$2.77 \times 10^{-3} \pm 2.62 \times 10^{-5}$ [bcde]
S0V0.25	0.000	0.250	4.81 [defg]	18.2 [g]	25.5 [gh]	44.9 [abc]	28.8 [cde]	$2.71 \times 10^{-3} \pm 6.29 \times 10^{-5}$ [bcde]
S0V0.36	0.000	0.360	4.49 [fg]	18.7 [g]	27.0 [gh]	41.8 [bc]	28.7 [cde]	$2.66 \times 10^{-3} \pm 2.62 \times 10^{-5}$ [de]
S0V0.52	0.000	0.520	4.44 [g]	19.0 [g]	28.2 [gh]	38.7 [c]	26.5 [de]	$2.68 \times 10^{-3} \pm 6.3 \times 10^{-6}$ [cde]
S0V0.75	0.000	0.750	4.39 [g]	20.0 [fg]	27.9 [gh]	38.1 [c]	26.2 [e]	$2.61 \times 10^{-3} \pm 2.76 \times 10^{-5}$ [e]
S1V0	1.00	0.000	4.74 [efg]	27.1 [f]	22.8 [h]	47.9 [ab]	28.8 [cde]	$2.84 \times 10^{-3} \pm 6.93 \times 10^{-5}$ [abcde]
S1.31V0.3	1.31	0.300	5.52 [abcd]	37.4 [e]	38.0 [ef]	40.3 [bc]	32.9 [abcd]	$2.74 \times 10^{-3} \pm 2.69 \times 10^{-5}$ [bcde]
S1.31V0.435	1.31	0.435	5.50 [abcd]	37.2 [e]	38.9 [de]	39.5 [c]	30.6 [bcde]	$2.70 \times 10^{-3} \pm 3.61 \times 10^{-5}$ [bcde]
S1.31V0.625	1.31	0.625	5.22 [cdef]	36.3 [e]	39.2 [de]	39.7 [c]	29.9 [bcde]	$2.83 \times 10^{-3} \pm 2.83 \times 10^{-6}$ [abcde]
S1.71V0	1.71	0.000	5.31 [abcde]	38.8 [de]	30.6 [fg]	44.8 [abc]	34.6 [abc]	$2.82 \times 10^{-3} \pm 9.40 \times 10^{-5}$ [abcde]
S2.24V0.3	2.24	0.300	5.86 [abc]	46.8 [c]	44.9 [bcde]	40.7 [bc]	35.1 [abc]	$2.88 \times 10^{-3} \pm 1.89 \times 10^{-4}$ [abcd]
S2.24V0.435	2.24	0.435	5.91 [abc]	46.9 [c]	46.6 [bcd]	42.3 [bc]	34.2 [abc]	$2.83 \times 10^{-3} \pm 9.7 \times 10^{-5}$ [abcde]
S2.24V0.625	2.24	0.625	5.81 [abc]	44.8 [cd]	47.8 [abc]	39.5 [c]	33.1 [abcd]	$2.77 \times 10^{-3} \pm 2.40 \times 10^{-5}$ [bcde]
S2.92V0	2.92	0.000	5.43 [abcde]	50.2 [c]	38.4 [e]	41.8 [bc]	35.2 [abc]	$2.85 \times 10^{-3} \pm 1.41 \times 10^{-6}$ [abcde]
S3.82V0.3	3.82	0.300	5.77 [abc]	58.8 [b]	50.3 [abc]	40.4 [bc]	35.4 [ab]	$2.93 \times 10^{-3} \pm 7.64 \times 10^{-5}$ [abc]
S3.82V0.435	3.82	0.435	6.00 [ab]	61.0 [ab]	51.9 [ab]	39.9 [bc]	35.4 [ab]	$2.96 \times 10^{-3} \pm 4.95 \times 10^{-5}$ [ab]
S3.82V0.625	3.82	0.625	6.04 [a]	58.1 [b]	54.7 [a]	37.8 [c]	36.4 [ab]	$2.94 \times 10^{-3} \pm 1.41 \times 10^{-6}$ [ab]
S5V0	5.00	0.000	5.27 [bcde]	67.7 [a]	43.4 [cde]	39.4 [c]	37.4 [a]	$3.06 \times 10^{-3} \pm 9.90 \times 10^{-6}$ [a]

Same superscript letters in columns do not differ by Tukey's HSD ($p < 0.05$).

References

1. Delwiche, J. The impact of perceptual interactions on perceived flavor. *Food Qual. Prefer.* **2004**, *15*, 137–146. [CrossRef]
2. Knoop, J.E. Cross-Modal Interactions in Complex Food Matrices. Ph.D. Thesis, Wageningen University, Wageningen, The Netherlands, November 2011.
3. Noble, A.C. Taste-aroma interactions. *Trends Food Sci. Technol.* **1996**, *7*, 439–444. [CrossRef]
4. Blake, A.A. Flavour Perception and the Learning of Food Preferences. In *Flavor Perception*; Taylor, A.J., Roberts, D.D., Eds.; Blackwell Publishing Ltd.: Oxford, UK, 2004; pp. 172–202.
5. Gilbert, A.N.; Firestein, S. Dollars and scents: Commercial opportunities in olfaction and taste. *Nat. Neurosci.* **2002**, *5*, 1043–1045. [CrossRef] [PubMed]
6. Shepherd, G.M. Smell images and the flavour system in the human brain. *Nature* **2006**, *444*, 316–321. [CrossRef] [PubMed]
7. Frank, R.A.; Byram, J. Taste-smell interactions are tastant and odorant dependent. *Chem. Senses* **1988**, *13*, 445–455. [CrossRef]
8. Köster, M.A.; Prescott, J.; Köster, E.P. Incidental learning and memory for three basic tastes in food. *Chem. Senses* **2004**, *29*, 441–453. [CrossRef] [PubMed]
9. Prescott, J. Flavour as a psychological construct: Implications for perceiving and measuring the sensory qualities of foods. *Food Qual. Prefer.* **1999**, *10*, 349–356. [CrossRef]
10. Frank, R.A.; van der Klaauw, N.J.; Schifferstein, H.N.J. Both perceptual and conceptual factors influence taste-odor and taste-taste interactions. *Percept. Psychophys.* **1993**, *54*, 343–354. [CrossRef] [PubMed]
11. Lawless, H.T.; Schlegel, M.P. Direct and indirect scaling of sensory differences in simple taste and odor mixtures. *J. Food Sci.* **1984**, *49*, 44–46. [CrossRef]
12. Schifferstein, H.N.J.; Verlegh, P.W.J. The role of congruency and pleasantness in odor-induced taste enhancement. *Acta Psychol.* **1996**, *94*, 87–105. [CrossRef]
13. Djordjevic, J.; Zatorre, R.J.; Jones-Gotman, M. Odor-induced changes in taste perception. *Exp. Brain Res.* **2004**, *159*, 405–408. [CrossRef] [PubMed]
14. Pfeiffer, J.C.; Hort, J.; Hollowood, T.A.; Taylor, A.J. Taste-Aroma interactions in a ternary system: A model of fruitiness perception in sucrose/acid solutions. *Percept. Psychophys.* **2006**, *68*, 216–227. [CrossRef] [PubMed]
15. Boakes, R.A.; Hemberger, H. Odour-modulation of taste ratings by chefs. *Food Qual. Prefer.* **2012**, *25*, 81–86. [CrossRef]
16. Stevenson, R.J.; Prescott, J.; Boakes, R.A. Confusing tastes and smells: How odours can influence the perception of sweet and sour tastes. *Chem. Senses* **1999**, *24*, 627–635. [CrossRef] [PubMed]
17. Labbe, D.; Rytz, A.; Morgenegg, C.; Ali, S.; Martin, N. Subthreshold olfactory stimulation can enhance sweetness. *Chem. Senses* **2007**, *32*, 205–214. [CrossRef] [PubMed]
18. Frank, R.A.; Ducheny, K.; Mize, S.J.S. Strawberry odor, but not red color, enhances the sweetness of sucrose solutions. *Chem. Senses* **1989**, *14*, 371–377. [CrossRef]
19. Stevenson, R.J. Is sweetness taste enhancement cognitively impenetrable? Effects of exposure, training and knowledge. *Appetite* **2001**, *36*, 241–242. [CrossRef] [PubMed]
20. Prescott, J.; Johnstone, V.; Francis, J. Odor-taste interactions: Effects of attentional strategies during exposure. *Chem. Senses* **2004**, *29*, 331–340. [CrossRef] [PubMed]
21. Bingham, A.F.; Birch, G.G.; De Graaf, C.; Behan, J.M.; Perring, K.D. Sensory studies with sucrose-maltol mixtures. *Chem. Senses* **1990**, *15*, 447–456. [CrossRef]
22. Arvisenet, G.; Guichard, E.; Ballester, J. Taste-aroma interaction in model wines: Effect of training and expertise. *Food Qual. Prefer.* **2016**, *52*, 211–221. [CrossRef]
23. Hewson, L.; Hollowood, T.; Chandra, S.; Hort, J. Taste-aroma interactions in a citrus flavoured model beverage system: Similarities and differences between acid and sugar type. *Food Qual. Prefer.* **2008**, *19*, 323–334. [CrossRef]
24. Jones, P.R.; Gawel, R.; Francis, I.L.; Waters, E.J. The influence of interactions between major white wine components on the aroma, flavour and texture of model white wine. *Food Qual. Prefer.* **2008**, *19*, 596–607. [CrossRef]
25. Lethuaut, L.; Weel, K.G.C.; Boelrijk, A.E.M.; Brossard, C.D. Flavor perception and aroma release from model dairy desserts. *J. Agric. Food Chem.* **2004**, *52*, 3478–3485. [CrossRef] [PubMed]

26. Bult, J.H.F.; De Wijk, R.A.; Hummel, T. Investigations on multimodal sensory integration: Texture, taste, and ortho- and retronasal olfactory stimuli in concert. *Neurosci. Lett.* **2007**, *411*, 6–10. [CrossRef] [PubMed]

27. Lavin, J.G.; Lawless, H.T. Effects of color and odor on judgments of Sweetness among children and adults. *Food Qual. Prefer.* **1998**, *9*, 283–289. [CrossRef]

28. Von Sydow, E.; Moskowitz, H.; Jacobs, H.; Meiselnan, H. Odor-taste interctions in fruit juices. *Leb. Technol.* **1974**, *7*, 18–24.

29. Symoneaux, R.; Guichard, H.; Le Quéré, J.-M.; Baron, A.; Chollet, S. Could cider aroma modify cider mouthfeel properties? *Food Qual. Prefer.* **2015**, *45*, 11–17. [CrossRef]

30. Green, B.G.; Nachtigal, D.; Hammond, S.; Lim, J. Enhancement of retronasal odors by taste. *Chem. Senses* **2012**, *37*, 77–86. [CrossRef] [PubMed]

31. Dalton, P.; Doolitle, N.; Nagata, H.; Breslin, P.A.S. The merging of the senses: Integration of subthreshold taste and smell. *Nat. Neurosci.* **2000**, 10–12. [CrossRef] [PubMed]

32. Valdes, R.M.; Simone, M.J.; Hinreiner, E.H. Effect of sucrose and organic acids on apparent flavor intensity. II. Fruit nectars. *Food Technol.* **1956**, *10*, 387–390.

33. Valdes, R.M.; Hinreiner, E.H.; Simon, M.J. Effect of sucrose and organic acids on apparent flavor intensity. I. Aqueous solutions. *Food Technol.* **1956**, *10*, 282–285.

34. Mintel US Dairy Milk Market Report. Available online: https://store.mintel.com/us-dairy-milk-market-report (accessed on 5 February 2018).

35. Havkin-Frenkel, D.; Belanger, F.C. *Handbook of Vanilla Science and Technology*; Wiley-Blackwell: Oxford, UK, 2010; ISBN 9781405193252.

36. Kuo, Y.-L.; Pangborn, R.M.; Noble, A.C. Temporal patterns of nasal, oral, and retronasal perception of citral and vanillin and interaction of these odourants with selected tastants. *Int. J. Food Sci. Technol.* **1993**, *28*, 127–137. [CrossRef]

37. Clark, C.C.; Lawless, H.T. Limiting response alternatives in time-intensity scaling: An examination of the halo-dumping effect. *Chem. Senses* **1994**, *19*, 583–594. [CrossRef] [PubMed]

38. Small, D.M.; Voss, J.; Mak, Y.E.; Simmons, K.B.; Parrish, T.; Dana, M.; Voss, J.; Mak, Y.E.; Simmons, K.B.; Parrish, T.; et al. Experience-dependent neural integration of taste and smell in the human brain. *J. Neurophsyiol.* **2004**, *92*, 1892–1903. [CrossRef] [PubMed]

39. Sakai, N. Enhancement of sweetness ratings of aspartame by a vanilla odor presented either by orthonasal or retronasal routes. *Percept. Mot. Skills* **2001**, *92*, 1002–1008. [CrossRef] [PubMed]

40. Fujimaru, T.; Lim, J. Effects of stimulus intensity on odor enhancement by taste. *Chemosens. Percept.* **2013**, *6*, 1–7. [CrossRef]

41. Lim, J.; Fujimaru, T.; Linscott, T.D. The role of congruency in taste-odor interactions. *Food Qual. Prefer.* **2014**, *34*, 5–13. [CrossRef]

42. Running, C.A.; Hayes, J.E. Sip and spit or sip and swallow: Choice of method differentially alters taste intensity estimates across stimuli. *Physiol. Behav.* **2017**, *181*, 95–99. [CrossRef] [PubMed]

43. Pizarek, A.; Vickers, Z. Effects of swallowing and spitting on flavor intensity. *J. Sens. Stud.* **2017**, *32*, 1–7. [CrossRef]

44. Suhnel, J. Evaluation of synergism or antagonism for the combined action of antiviral agents. *Antiviral Res.* **1990**, *13*, 23–40. [CrossRef]

45. Suhnel, J. Zero interaction response surfaces, interaction functions and difference response surfaces for combinations of biologically active agents. *Arzneim. Forsch.* **1992**, *42*, 1251–1258.

46. Tallarida, R.J. Revisiting the isobole and related quantitative methods for assessing drug synergism. *J. Pharmacol. Exp. Ther.* **2012**, *342*, 2–8. [CrossRef] [PubMed]

47. Berenbaum, M.C. Synergy, additivism and antagonism in immunosuppression: A critical review. *Clin. Exp. Immunol.* **1977**, *28*, 1–18. [PubMed]

48. Berenbaum, M.C. What is synergy? *Pharmacol. Rev.* **1989**, *41*, 93–141. [PubMed]

49. Fleming, E.E.; Ziegler, G.R.; Hayes, J.E. Investigating mixture interactions of astringent stimuli using the isobole approach. *Chem. Senses* **2016**, *41*, 601–610. [CrossRef] [PubMed]

50. Wolf, P.A.; Bridges, J.R.; Wicklund, R. Application of agonist-receptor modeling to the sweetness synergy between high fructose corn syrup and sucralose, and between high-potency sweeteners. *J. Food Sci.* **2010**, *75*. [CrossRef] [PubMed]

51. Reyes, M.M. Using Psychophysical, Physicohedonic, and Consumer Choice Responses to Sweetener Technology to Support Milk Intake in Adolescent Women. Ph.D. Thesis, The Pennsylvania State University, State College, PA, USA, 2017.

52. Chandan, R.C. *Manufacturing Yogurt and Fermented Milks*; Blackwell Publishing: Victoria, Australia, 2006; ISBN 9780813823041.

53. Lawless, H.T.; Heymann, H. Acceptance Testing. In *Sensory Evaluation of Food*; Lawless, H.T., Heymann, H., Eds.; Springer: New York, NY, USA, 2010; pp. 325–347.

54. R Core Team. *R: A Langugage and Environment for Statistical Computing*; R Foundation for Statistical Computing: Vienna, Austria, 2018; ISBN 3900051070. Available online: http://www.r-project.org (accessed on 1 February 2018).

55. De Mendiburu, F. Package "Agricolae". Available online: https://cran.r-project.org/package=agricolae (accessed on 5 February 2018).

56. Lenth, R. Package "rsm". Available online: https://cran.r-project.org/package=rsm (accessed on 5 February 2018).

57. Hayes, J.E.; Duffy, V.B. Oral sensory phenotype identifies level of sugar and fat required for maximal liking. *Physiol. Behav.* **2008**, *95*, 77–87. [CrossRef] [PubMed]

58. Suhnel, J. Evaluation of interaction in olfactory and taste mixtures. *Chem. Senses* **1993**, *18*, 131–149. [CrossRef]

59. Grab, W. Blended Flavourings. In *Flavourings: Production, Composition, Applications, Regulations*; Ziegler, H., Ed.; Wiley-VCH: Weinheim, Germany, 2007; pp. 391–434. ISBN 9783527314065.

60. Li, B.; Hayes, J.E.; Ziegler, G.R. Just-About-Right and ideal scaling provide similar insights into the influence of sensory attributes on liking. *Food Qual. Prefer.* **2014**, *37*, 71–78. [CrossRef] [PubMed]

61. Welge-Lüssen, A.; Husner, A.; Wolfensberger, M.; Hummel, T. Influence of simultaneous gustatory stimuli on orthonasal and retronasal olfaction. *Neurosci. Lett.* **2009**, *454*, 124–128. [CrossRef] [PubMed]

62. Alcaire, F.; Lucia, A.; Vidal, L.; Gimenez, A.; Ares, G. Aroma-related cross-modal interactions for sugar reduction in milk desserts: Influence on consumer perception. *Food Res. Int.* **2017**, *97*, 45–50. [CrossRef] [PubMed]

63. Linscott, T.D.; Lim, J. Retronasal odor enhancement by salty and umami tastes. *Food Qual. Prefer.* **2016**, *48*, 1–10. [CrossRef]

64. Cayeux, I.; Mercier, C. Sensory evaluation of interaction between smell and taste-application to sourness. In *Flavour Research at the Dawn of the Twenty-First Century*; Le Quéré, J.-M., Etievant, P.X., Eds., Editions Tec & Doc: Paris, France, 2003; pp. 287–292.

65. Isogai, T.; Wise, P.M. The effects of odor quality and temporal asynchrony on modulation of taste intensity by retronasal odor. *Chem. Senses* **2016**, *41*, 557–566. [CrossRef] [PubMed]

66. Dravnieks, A. *Atlas of Odor Character Profiles*; ASTM: Philadelphia, PA, USA, 1985.

67. Stevenson, R.J.; Prescott, J.; Boakes, R.A. The acquisition of taste properties by odors. *Learn. Motiv.* **1995**, *26*, 433–455. [CrossRef]

68. Bonnans, S.; Noble, A.C. Effect of sweetener type and of sweetener and acid levels on temporal perception of sweetness, sourness and fruitiness. *Chem. Senses* **1993**, *18*, 273–283. [CrossRef]

69. Valentin, D.; Chrea, C.; Nguyen, D.H. Taste-odour interactions in sweet taste perception. In *Optimising Sweet Taste in Foods Foods*; Woodhead Publishing Ltd.: Cambridge, UK, 2006; pp. 1–18.

70. Djordjevic, J.; Zatorre, R.J.; Jones-Gotman, M. Effects of perceived and imagined odors on taste detection. *Chem. Senses* **2004**, *29*, 199–208. [CrossRef] [PubMed]

71. Nguyen, D.H.; Dacremont, C.; Valentin, D. Taste-odour interactions and perceptual separability. *Vietnam J. Sci. Technol.* **2011**, *49*, 73–82.

72. Brossard, C.D.; Lethuaut, L.; Boelrijk, A.E.M.; Mariette, F.; Genot, C. Sweetness and aroma perceptions in model dairy desserts: An overview. *Flavour Fragr. J.* **2006**, *21*, 48–52. [CrossRef]

73. Small, D.M.; Jones-Gotman, M.K. Neural substrates of taste/smell interaction and flavor in the human brain. *Chem. Senses* **2001**, *26*, 1034.

74. De Araujo, I.E.T.; Rolls, E.T.; Kringelbach, M.L.; Mcglone, F.; Phillips, N. Taste-olfactory convergence, and the representation of the pleasantness of flavour, in the human brain. *Eur. J. Neurosci.* **2003**, *18*, 2059–2068. [CrossRef] [PubMed]

75. Kuznicki, J.T.; Hayward, M.; Schultz, J. Perceptual processing of taste quality. *Chem. Senses* **1983**, *7*, 273–292. [CrossRef]

76. McBurney, D.H. Taste, smell, and flavor terminology: Taking the confusion out of fusion. In *Clinical Measurement of Taste and Smell*; Meiselman, H.L., Rivkin, R.S., Eds.; MacMillan: New York, NY, USA, 1986; pp. 117–125.

77. Van der Klaauw, N.J.; Frank, R.A. Matching and scaling of taste-smell mixtures: Individual differences in sweetness enhancement by strawberry odor. *Chem. Senses* **1994**, *19*, 567.

Article

Sensory Impact of Polyphenolic Composition on the Oxidative Notes of Chardonnay Wines

Jordi Ballester [1],*, Mathilde Magne [2], Perrine Julien [1] Laurence Noret [2], Maria Nikolantonaki [2], Christian Coelho [2] and Régis D. Gougeon [2]

[1] Centre des Sciences du Goût et de l'Alimentation, UMR 6265 CNRS, UMR 1324 INRA-Université de Bourgogne Franche Comté, 9 E Boulevard Jeanne d'Arc, F-21000 Dijon, France; perrine.julien@laposte.net

[2] UMR A 02.102 PAM, Université de Bourgogne Franche Comté, Institut Universitaire de la vigne et du vin Jules Guyot, rue Claude Ladrey, BP 27877, 21078 Dijon CEDEX, France; mathilde.magne@agroparistech.fr (M.M.); laurence.noret@u-bourgogne.fr (L.N.); maria.nikolantonaki@u-bourgogne.fr (M.N.); christian.coelho@u-bourgogne.fr (C.C.); regis.gougeon@u-bourgogne.fr (R.D.G.)

* Correspondence: jordi.ballester@u-bourgogne.fr; Tel.: +33-380-396-393

Academic Editors: Manuel Malfeito Ferreira and Richard Owusu-Apenten
Received: 29 November 2017; Accepted: 1 February 2018; Published: 1 March 2018

Abstract: Chardonnay wines have a long-standing reputation regarding their aging potential. However, in some cases, they face premature oxidation a few years after bottling. Scientific reports are, for now, multiparametric and unclear. Polyphenols seem to be an important factor involved in the oxidative stability of white wines, but their role has not yet been completely characterized. The present study aimed to investigate the link between polyphenol content and the emergence of oxidative odors of bottle-aged Chardonnay wines. In order to obtain samples with noticeable differences in polyphenol content, as well as in sensory oxidative notes, wines from two different vintages were used. For each vintage, three levels of must clarification and two wine closures were implemented. Polyphenol content was analyzed chemically, and the oxidative character was assessed sensorially by a trained panel using a specific intensity scale. The results showed significant effects for closure type and turbidity. However, these effects were strongly affected by vintage. Concerning the polyphenol content, a clear difference was also found between vintages, closures and turbidity levels. Significant linear regression models for REDOX scores pointed out Flavon-3-ols as the main negative predictor, and grape reaction product (GRP) as the main positive predictor. The enological implications are discussed.

Keywords: flavan-3-ols; reduction; oxidation; wine aging; oxidative stability; clarification

---

1. Introduction

Chardonnay wines have a long-standing reputation regarding their aging potential. During bottle aging, wine is exposed to relatively low quantities of oxygen, which are nevertheless sufficient to influence its sensory characteristics [1]. In particular, oxygen modulates the extent of different reactions involving volatile and nonvolatile components, resulting in the formation/degradation of a number of powerful aroma compounds, with major consequences for the process of aroma evolution during bottle aging [2]. In addition, other chemical reactions taking place during bottle aging do not involve oxygen, meaning that, even in an environment completely devoid of oxygen, a certain form of aging will occur [3]. Wine aroma changes dramatically during bottle aging, through a complex array of chemical reactions that are only partly understood. In most cases, oxygen and polyphenols contribute to the evolution of these key aroma compounds [4] where iron and copper act as oxidation catalyzers [5]. To date, the majority of studies dealing with premature oxidation have focused on the characterization of potent oxidation markers of defective aroma (off-flavors). According to these studies, changes in wine aroma properties linked to oxidation were related to the formation of off-flavors, mainly aldehyde compounds, such as phenylacetaldehyde, methional, *trans*-2-nonenal, o-aminoacetophenone and a lactone, the 3-hydroxy-4,5-dimethyl-2(5*H*)-furanone (sotolon) [6,7]. The aldehydes, methional and phenylacetaldehyde, are among the oxidation-related aroma compounds that have drawn the most attention, due to their supposedly higher aroma impact and possible contribution to the aroma of red and white wines [8]. For these aldehydes, formation from the amino acids methionine and phenylalanine, respectively, via Strecker reaction involving the presence of a dicarbonyl compound has been proposed [9]. Different o-diphenols have been shown to form different quantities of aldehydes, with caffeic acid giving higher methional and phenylacetaldehyde compared to catechin and epicatechin [10], but this was observed at pH much higher than that of wine. More recently, it has been shown that in wine-like conditions, methionine and phenylalanine were not capable of reacting with a model quinone [11]. Finally, in a recent study, the polyfunctional thiol 3-ethylsulfanyl acetate was identified for the first time as a key contributor to off-flavors in sauvignon blanc wines [12]. This compound was observed at higher levels in aged wines and in wines obtained from juices exposed to air, but the effect of post-bottling oxygen exposure on its concentration remains to be investigated. Overall, although many aroma compounds relevant to wine oxidation have been identified, most studies have been aimed at understanding wine oxidative spoilage, and have often been carried out under conditions of extreme oxygen exposure. Conversely, oxidative processes taking place during bottle aging under normal conditions are rather "mild", and the sensory impact of such levels of oxidation to the aroma quality of wines remains to be established.

Many chemical approaches have been developed in the literature to understand white wine oxidation. However, there are fewer studies dealing with white wine oxidation from a sensory point of view. Typically, the researchers use a tasting panel to generate relevant attributes, which usually cover fresh fruit dimensions (depending on the grape variety at hand), oxidative aroma notes and a number of other descriptors unrelated with oxidation (such as spiciness, oakyness and other wine faults). Several authors have used descriptive analysis to characterize the type of closure and the effect of ascorbic acid addition on the oxidative or reductive aromas of white wine samples with several years of storage [13–15]. In Gooden et al. [13], fresh fruit attributes were overall fruit, pineapple, citrus/lime and tropical while the oxidation related attributes were oxidized and developed. Interestingly, the authors aimed to differentiate two different nuances of oxidation, developed being probably a mild and not necessarily negative version of oxidized. As expected, the authors showed a clear opposition between all fresh fruit dimensions, and oxidized and developed. Ulterior studies [14–16] slightly modified the lexicon by including some oxidation notes such as aldehyde, glue-like or wet wool, and some reduction notes, such as struck flint/rubber, gunflint, rotten egg, cabbage or stagnant water. In the case of rosé wine, Guaita and coworkers [17] suggested, despite the lack of significance of the results, that the closures with higher Oxygen Transfer Rate (OTR) values induced a stronger intensity of acacia flower and rose aromas in the wines than more protective closures. In the same vein, Rodrigues and coworkers [18] used the attribute honey/wax as an oxidation marker. A different strategy was used by Brajkovich and coworkers [19], since they basically used sauvignon blanc varietal descriptors such as passion fruit, cat urine, grassy and capsicum, and showed that the most oxidized samples were those showing lower intensities for thiol-related attributes.

In that context, the goal of the present study is to investigate the oxidative note of bottle-aged Chardonnay wines from a polyphenolic content perspective. To the best of our knowledge, no sensory studies have so far examined the link between the occurrence of oxidative aromas and the polyphenol content in white wines, in particular hydroxycinnamic acids and flavanols. This is all the more interesting because previous studies reported contradictory results; some have considered that the antioxidant activity of white wines is related to its polyphenolic content [20–22], whereas others have shown a direct correlation between the browning rate and the concentration of epicatechin [23]. In order to ensure wide differences in the polyphenol concentrations and sensory characteristics between the samples, two different vintages with three levels of must turbidity and two different types of closure were used. The two kinds of stoppers chosen for this study (synthetic coextruded stoppers and screw cap) presented extreme OTR values to ensure oxidation differences as previously described [3]. Synthetic coextruded stoppers are able to diffuse 10 to 100 times more oxygen that screw caps, and could possibly reveal oxidative deviations earlier in wine compared to more reductive environments when screw caps are used [3,24].

2. Materials and Methods

2.1. Wines

Chardonnay dry white wines from Burgundy were elaborated following the same winemaking process during vintages 2009 and 2010. Chardonnay grapes were hand-harvested and pressed with a pneumatic press. Must was protected with 4 g·hL^{-1} of SO_2. Cold racking at 12 °C from 12 to 24 h enabled to reach the three levels of must turbidity: 300 NTU (Low), 600 NTU (Medium) and 800 NTU (High). Alcoholic and malolactic fermentations were conducted in oak barrels followed by oak aging for 6 months. Dry white wines were then filtered, and SO_2 was adjusted to 40 mg·L^{-1} prior to bottling. Wine bottles were stored in a cellar with a constant temperature (12 °C) until required for chemical analysis (April 2015). Table 1 summarizes the characteristics of the samples used in this study.

Table 1. Characteristics of the white wines used in this study and their codes.

Sample Code	Closure	Turbidity (NTU)
S-L-2009	synthetic coextruded stopper	LOW
S-M-2009	synthetic coextruded stopper	MEDIUM
S-H-2009	synthetic coextruded stopper	HIGH
C-L-2009	screw cap	LOW
C-M-2009	screw cap	MEDIUM
C-H-2009	screw cap	HIGH
S-L-2010	synthetic coextruded stopper	LOW
S-M-2010	synthetic coextruded stopper	MEDIUM
S-H-2010	synthetic coextruded stopper	HIGH
C-L-2010	screw cap	LOW
C-M-2010	screw cap	MEDIUM
C-H-2010	screw cap	HIGH

2.2. Sensory Analysis

Sensory sessions took place between October and December 2015. After a training period, the selected assessors performed a monadic assessment of the reductive and oxidative aromas of the samples and a sensory description on frequency of citation [25]. Afterwards, a subset of the samples was analyzed again, not in monadic presentation, but in simultaneous presentations in order to induce a comparative assessment and therefore detect more subtle differences between samples.

2.2.1. Panel: Training and Selection

Twenty-six candidates were recruited from among the students at the enology school of Dijon (France). As part of their enology training, they all attended a wine-tasting course (36 h), where they learned the main olfactory notes of wine. Apart from this general training, the panelists took part in 6 training sessions and 4 selection sessions organized as part of the present study. The general goal of the training was to familiarize the assessors to the common oxidation and reduction aroma notes, and to quantitatively calibrate their measurements. With this purpose, natural and spiked wines with a range of reduction or oxidation intensities were selected for the training. Also, several "clean" wines (with no obvious oxidation or reduction notes) were used as controls during the training. Sessions were conducted twice a week, in groups of 10–13 candidates and lasted about 40 min. The specific goals and the content of each session are summarized in Table 2. A specific structured scale (called here the REDOX odor scale) was created to assess the global oxidative or reductive odor of the samples. This 11-point scale went from -5 (strong reduction) to +5 (strong oxidation), with 0 (neither reduced nor oxidized) being in the middle of the scale. Expected REDOX odor levels of the training samples were roughly determined by consensus by three experienced tasters as a guideline only. Systematic feedback was given at the end of each session or at the beginning of the subsequent session.

Table 2. Summary of the training and selection sessions prior to sensory description.

Session Number	Goals and Protocols	Materials
Session 1	Familiarization with phenylacetaldehyde and methional in flasks and in water. Furaneol (oaky note) was used as a distractor. Assessors' descriptions were discussed and compared to the descriptors from the literature.	Phenylacetaldehyde 2,2 µg/L Methional 5 mg/L Furfural, one drop on a cotton ball in a flask.
Session 2	Familiarization with sotolon, 2-aminoacetophenone and methional in flasks and in water. *Trans*-2-nonanal (cardboard note) was used as a distractor. Assessors' descriptions were discussed and compared to the descriptors from the literature.	Sotolon 90 µg/L 2-aminoacetophenone 10 µg/L Methional 5 mg/L *Trans*-2-nonanal, one drop on a cotton ball in a flask.
Session 3	Familiarization with reduction notes: H2S, Ethanthiol, DMS. Recognition of 2-aminoacetophenone and phenylacetaldehyde. Assessors' descriptions were discussed and compared to the descriptors from the literature. Familiarization with the redox scale on the previous solutions	H2S 40 µg/L Ethanthiol 200 µg/L DMS 55 µg/L 2-aminoacetophenone 10 µg/L Phenylacetaldehyde 14 µg/L
Session 4	Discrimination between oxidation and reduction notes in spiked wines and recognition of the different molecules and their descriptors. Familiarization with the redox scale on the previous spiked wines	Base wine (Muscadet, 2014) Base wine + phenylacetaldehyde 14 µg/L Base wine + methional 5 mg/L Base wine + Ethanthiol 200 µg/L Base wine + sotolon 120 µg/L Base wine + 2-aminoacétophenone 10 µg/L
Session 5	Test assessors' discrimination abilities on oxidized samples (spiked wines) by means of a triangle test. Odor characterization of 3 samples using the redox scale and the odor descriptors list.	Base wine (Muscadet, 2014) Used in the triangle test: Base wine + phenylacetaldehyde 2.2 µg/L + sotolon 90 µg/L and Base wine + 2-aminoacétophenone 15 µg/L + sotolon 90 µg/L
Session 6	Test assessors' discrimination abilities between an oxidized sample (spiked wines) and the base wine. Odor characterization of 2 samples using the redox scale and the odor descriptors list.	Base wine (Muscadet, 2014) Base wine + sotolon 90 µg/L

Table 2. *Cont.*

Session Number	Goals and Protocols	Materials
Session 7	Quantitative calibration of the anchors of the scale. Redox rating practice with the redox scale on three wines previously checked by two wine experts as representative of three oxidation levels.	Used for calibration: Reduction anchor (very reduced): H_2S 40 µg/L Oxidation anchor (very oxidized): St Aubin 1998 in a dark ISO glass Used for practice with the redox scale: Saint Aubin 1998 (very oxidized) Marsannay 2007 (moderately oxidized) Viré Clessé 2014 (neither reduced nor oxidized)
Session 8	Redox rating of 4 wines previously checked by two wine experts as representative of three oxidation levels.	Bourgogne Aligoté 2014 (neither reduced nor oxidized) Chardonnay blend with 7% of St Aubin 1998 (slightly oxidized) Petit Chablis 2009 (moderately oxidized) Chablis 1er Cru 2012 (slightly reduced)
Session 9	Redox rating of 6 wines previously checked by two wine experts as representative of three oxidation levels. The last three samples are tasted in session 10 as well in order to check candidates' repeatability.	Chassagne Montrachet, 2002 (moderately oxidized) Chenin Blanc 2013 (slightly reduced) Mâcon Clessé 1994 (very oxidized) Viré Clessé 2014 + Petit Chablis 2009 (moderately oxidized) Chardonnay Pays d'Oc 2014 (clean or just slightly oxidized) Chardonnay Pays d'Oc 2014+ 20% Saint Aubin 1er Cru 1998 (moderately oxidized) Chardonnay California 2010 (moderately oxidized)
Session 10	Redox rating of 6 wines previously checked by two wine experts as representative of three oxidation levels. The last three samples are tasted in session 10 as well, in order to check candidates' repeatability.	Chablis 1er Cru 2012 Aligoté Marsannay 2015 (neither reduced nor oxidized) Viré-Clessé 2014 + 20% Ugni Blanc 1969 (moderately oxidized) Viognier 2013 + 50% Chardonnay 2013 (clean or just slightly oxidized) Chardonnay Pays d'Oc 2014 (clean or just slightly oxidized) Chardonnay Pays d'Oc 2014+ 20% SaintAubin 1er Cru 1998 (moderately oxidized) Chardonnay California 2010 (moderately oxidized)

The selection of the panelists was based on the results of sessions 7 to 10. Three main criteria were considered in order to rank the candidates according to their sensory performance: recognition of oxidation and reduction notes and proximity with the expected REDOX odor levels, repeatability, and consensus with the rest of the panel. Panelists were ranked according to each of these three criteria, and then an average rank was computed for each candidate.

At the end of the selection process, 14 assessors were selected to form the final panel (three females and eleven males, average age 24.4; SD = 1.4). Among them, eleven (three females and eight males, average age 24.3; SD = 1.3) participated in the monadic sensory sessions. All fourteen assessors participated in the comparative REDOX odor assessment.

2.2.2. Monadic Sensory Description

Sensory analyses took place in a sensory room equipped with individual booths. Two bottles of each sample were assessed in standardized black glasses coded by 3-digit numbers. Each sample contained 20 mL of wine at ambient temperature and was covered with a plastic Petri dish. First, participants were asked to rate the intensity of oxidation-reduction of the samples orthonasally (i.e., by smell only) and afterwards globally (nose and palate) using the REDOX odor scale. Finally, panelists were asked to describe the odor of the samples using a list of odor attributes (Figure 1) from which assessors should select all the attributes they considered appropriate to describe each sample. The sensory terms aimed to cover varietal chardonnay aromas, oxidation and reduction notes and other wine faults (in order to control for false positives). The list was based on previous research on white wine oxidation [13–16], and the terms were listed in alphabetical order. If panelists needed an attribute that was missing from the list, two slots (other 1 and other 2) were allowed for their own descriptors. During each session, the twelve samples (six wines replicated) were randomly split into three series of four samples. The tasting order within the series was specific to each participant and followed a William's Latin square. In order to reduce sensory fatigue, panelists were asked to take a short break between series. Monadic sensory description took two sessions, one for 2009 samples and another, the day after, for the 2010 samples.

Bitter almond		Herbaceous		Rotten egg/cabbage	
Bruised apple		Honey		Spicy	
Butter		Leather/Stable		Stagnant water	
Caramel/vanila		Mineral		Toasted	
Citrus		nail polish remover		Tropical fruits	
Cooked vegetables		Woody		Wallnut/curry	
Corked		Prune		Wax/mothball	
Dust/carboard		Quince		White fruits	
Floral		Rancid butter		Yellow fruits	
Forest floor		Rancid honey		Other 1	
Hay		Rancio/madere		Other2	

Figure 1. Ballot used for the general odor description (translated from French).

2.2.3. Comparative REDOX Odor Assessment between Turbidity Levels with Synthetic Coextruded Stopper

This second descriptive analysis was carried out in order to focus specifically on the effect of turbidity. According to the results of the monadic description (see Section 3.1), the samples closed with screw caps barely showed oxidative notes. In particular, for the 2010 samples, the differences in redox score between closure types were really important which were assumed to mask subtle differences between turbidity levels (see Section 3.1). Therefore, only the samples with synthetic coextruded stoppers were chosen for the comparative REDOX odor assessment. Moreover, inspired by Godden and coworkers [13], a full comparative approach was carried out in order to accede to more subtle differences.

Only one bottle of each treatment was used in this session. Comparative REDOX odor assessment took only one session, during which 6 samples were assessed. The three 2009 samples were simultaneously presented to the assessors (instead of monadically), in order to facilitate the comparison between samples. Panelists were allowed to freely retaste and compare the samples before scoring them. Then, after a short break, the three 2010 samples were presented for assessment, also simultaneously. As in the monadic descriptive analysis, participants were asked to rate the REDOX odor character of the samples orthonasally and, afterwards, globally, using the REDOX odor scale. In order to increase the sensitivity of the REDOX odor scale, a continuous version of the scale was presented to the panelists (instead the discrete scale used in the monadic profile). A subsequent general description based on frequencies of citation was not carried out in this case. Within each series, samples were presented according to a specific order for each participant following a William's Latin square.

2.3. Chemical Analysis

2.3.1. Reagents

Gallic acid, protocatechuic acid, hydroxybenzoic acid, tyrosol, hydroxytyrosol, salicylic acid, coumaric acid, ferulic acid, caffeic acid, caftaric acid, catechin, epicatehin, chlorogenic acid and gentisic acid were purchased from Sigma Aldrich. High-grade quality methanol and formic acid were used for chromatographic elutions.

2.3.2. Wine Polyphenols Analyses

Total polyphenols Index (TPI) measurements. Each wine sample was characterized by measuring the absorbance at 280 nm with a spectrophotometer (UV-1800, Shimadzu, Kyoto, Japan), after dilution, representing the global content of wine polyphenols expressed as TPI index.

UPLC-DAD/FLD phenolic compounds analysis. An Acquity UPLC H-Class (Waters, Milford, MA, USA) with a quaternary pump and an autosampler were coupled to a fluorimetric and a diode array detector. Under optimized conditions, the column oven was thermostated at 35 °C, and the sample system at 12 °C. The acquisitions, the solvent delivery and the detection were performed by Empower 2. About 2 mL of wine was filtered through a 0.45 μm PTFE filter (Restek, Lisses, France), of which 2 μL was injected into a reversed-phase BEH C18 (150 mm $\times$ 2.1 mm, 1.7 μm) Waters column. A binary solvent was run at a flow rate of 0.25 mL/min employing (A) H_2O:methanol (95:5) v/v formic acid (0.1%) and (B) methanol (100%). The optimized elution system consisted of a stepwise gradient as follows: from 3 to 5% B (0–4 min), 5 to 8% B (4–10 min), 8% B (10–12 min), 8 to 10% B (12–14 min), 10 to 15% B (14–17 min), 15 to 30.1% B (17–19 min), 30.1 to 38% B (19–21 min), 38 to 41% B (21–24 min), 41 to 50% B (24–30 min), 50 to 100% B (30–31 min), 100% B (31– 31.5 min), 100 to 3% B (31.5–32.5 min), 3% B (32.5–35 min). The diode array detector was set at 320 nm for *trans*-caftaric acid, gentisic acid, *trans*-caffeic acid, *trans*-coutaric acid, 2-S-glutathionylcaftaric acid (GRP), *trans*-ferulic acid, at 305 nm for salicylic acid, *trans*-coumaric acid, at 280 nm for gallic acid and at 260 nm for protocatechuic acid and hydroxybenzoïc acid. The fluorescence detector was set at λex = 270 nm and λem = 322 nm for hydroxytyrosol, tyrosol, catechin, epicatechin, proanthocyanidin B1, proanthocyanidin B2. Polyphenols were identified using a combination of commercial standards and the UV-visible spectra associated to chromatographic peaks in comparison with published procedures, as described previously [26,27]. *Trans*-Coutaric acid and GRP were quantified via their respective absorbance at 320 nm and expressed in *trans*-caftaric acid equivalents. Unknown concentrations were determined from the regression equation, and the results were converted into milligrams per liter. The sum of each individual polyphenol concentration was calculated and defined as the total polyphenol for five specific chemical families: phenolic acids, cinnamic acids, flavan-3-ols, Grape reaction product (GRP) and tyrosol.

2.4. Statistical Analysis

Differences in REDOX scores were tested statistically using analysis of variance (ANOVA) with $\alpha = 5\%$. Monadic data were subjected to three-way ANOVAs with wine closures (synthetic cork or screw cap) and turbidity (LOW, MEDIUM and HIGH) as within-subject factors. Subjects (panelists) were considered a random factor. Turbidity $\times$ closure, panelists $\times$ closure, and panelists $\times$ turbidity interactions were also tested. Main treatment effects (closure and turbidity) were tested using their respective interaction with panelists as the appropriate error term.

When a significant main effect was found, pairwise comparisons were carried out using Newman-Keuls' test ($\alpha = 5\%$). In the particular case of the comparative REDOX odor assessment, since all the samples had the same closure system, only the turbidity was set as a within-subject factor.

Concerning the general description, the frequency of citation of each term was determined for each wine. Only the descriptors cited at least 3 times for a given wine were considered in subsequent analyses; the other descriptors were discarded. The resulting contingency table containing the frequency of citation of each term for each wine was submitted to Correspondence Analysis (CA). In order to identify wine clusters, wine coordinates on the two first factors (F1 and F2) on both CAs were submitted to a Hierarchical Cluster Analysis (HCA) with the Ward criterion. Orthonasal and global REDOX average scores were projected as supplementary variables in each CA.

The effect of vintage, closure and turbidity on wine polyphenolic content was analyzed statistically by a three-factor ANOVA with ($\alpha = 5\%$) for each phenolic family. All second-order interactions were also included in the ANOVA model.

Relationships between polyphenol concentrations and REDOX odor scores were explored by means of multiple linear regressions (stepwise method), separately for 2009 and 2010. The added concentration of each family of polyphenols was used as predictors and the REDOX average intensity as the dependent variable.

All statistical analyses were carried out using XLStat 2017 (Addinsoft, Paris, France).

3. Results

3.1. Sensory Characterization by Monadic Profile

3.1.1. REDOX Odor Assessment

The results of the ANOVA performed on the orthonasal and global REDOX odor scores for 2009 and 2010 samples are presented in Table 3. The factor panelists showed a significant effect in both conditions and for both vintages. This is a common phenomenon in sensory evaluation, and indicates that the assessors used different parts of the scale.

Type of closure did not show significant effects for 2009 samples. Concerning the 2010 samples, significant panelists $\times$ closure interactions were found for both for orthonasal and global assessments, that some of the panelists did not achieve perfect concept alignment. The MS of panelists $\times$ closure interactions was used as an error term to test the closure effect, which was still significant for both assessment conditions (Table 3).

Turbidity level was significant for both conditions of the 2009 samples, but not for the 2010 ones. However, the interaction turbidity $\times$ closure was also significant, which means that it is not possible to generalize the turbidity effect to all the closure types. A closer look at this interaction (Figure 2a,b) shows that, for screw caps, medium and high turbidity levels tend to show lower REDOX odor values. However, this pattern does not fit the results for synthetic corks, for which the lowest turbidity also shows the lowest REDOX odor score (Figure 2a,b). Further research needs to be carried out to better understand the relationship between turbidity and oxidation.

Table 4 shows the results of the post-hoc mean comparison for Closure type and Turbidity level. Wines closed with synthetic cork showed significantly higher sensory oxidation than wines closed with screw caps, but only for 2010 samples.

Table 3. Summary of the significance level of the ANOVAS performed on the orthonasal and global REDOX odor scores for the 2009 and 2010 samples. Probabilities in bold are significant at $\alpha = 5\%$.

Assessment Condition	Sources of Variation	2009 Vintage		2010 Vintage	
		F	p	F	p
OTHONASAL PERCEPTION	PANELISTS	3.748	0.0003	3.633	0.0005
	CLOSURE	0.697	0.171	10.108	0.010
	TURBIDITY	3.412	0.038	0.540	0.591
	PANELISTS ×CLOSURE	3.122	0.002	3.289	0.001
	PANELISTS ×TURBIDITY	0.620	0.887	1.016	0.453
	TURBIDITY ×CLOSURE	5.880	0.004	2.081	0.131
GLOBAL PERCEPTION	PANELISTS	3.760	0.0003	3.812	0.0003
	CLOSURE	3.205	0.104	8.109	0.0173
	TURBIDITY	7.044	0.005	1.574	0.2318
	PANELISTS ×CLOSURE	1.827	0.068	2.870	0.0039
	PANELISTS ×TURBIDITY	0.683	0.832	0.994	0.4779
	TURBIDITY ×CLOSURE	3.182	0.046	0.287	0.7510

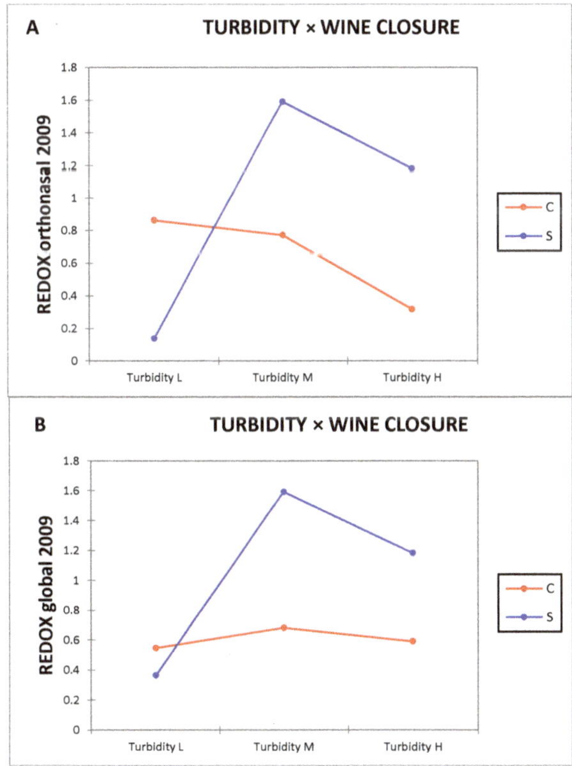

Figure 2. Turbidity*wine closure interaction plots for 2009 data (**A**) orthonasal evaluation and (**B**): global evaluation). S: synthetic coextruded stoppers; C: screw cap stoppers. L: low; M; medium; H: high.

Table 4. Results of the post-hoc Newman-Keuls' tests (α = 0.05) on the REDOX odor intensities comparing the two closure types and the three turbidity levels for each vintage and each evaluation condition. S: synthetic coextruded stoppers; C: screw cap stoppers. L: low; M: medium; H: high.

Factor	Level	2009 Orthonasal	2009 Global	2010 Orthonasal	2010 Global
CLOSURE	S	0.97 a	1.045 a	1.62 a	1.35 a
	C	0.65 a	0.61 a	0.38 b	0.53 b
TURBIDITY	M	1.182 a	1.14 a	1.16 a	1.045 a
	H	0.75 ab	0.89 ab	0.93 a	1.045 a
	L	0.5 b	0.45 b	0.91 a	0.73 a

Samples associated with the same letter were not significantly different.

Concerning the effect of turbidity and regardless the assessment condition, the REDOX odor score for the lowest level was significantly lower than the intermediate level, but wasn't different from the highest level. However, the significant closure*turbidity interactions found for 2009 samples prevent us from generalizing the significant differences found for turbidity.

3.1.2. General Description

Correspondence Analyses were carried out on the matrix containing the frequencies of citation of the most cited attributes for each vintage separately. Moreover, the average orthonasal and global REDOX odor scores were projected as supplementary variables, which were significantly correlated for both 2009 (r = 0.88, p < 0.05) and 2010 (r = 0.86, p < 0.05) vintages.

Concerning 2009 samples the CA followed by a HCA did not yield any meaningful or interpretable clusters (data not given). The four emerging clusters were not based on type of closure or turbidity level. Neither the samples nor the attributes seemed organized across a reduction–oxidation gradient (data not given).

Figure 3 shows the first and second dimensions of the CA on the frequencies of citation of the descriptors for the 2010 samples. Dimensions 1 and 2 accounted for 38.33% and 18.54% of the inertia, respectively.

According to cluster analysis, samples were sorted into two main groups. A first group, located in the negative values of the first dimension, was composed by samples with synthetic corks, and was described with oxidative terms such as "rancid honey", "cooked vegetables", "walnut/curry" and "bruised apple". Wines with screw caps were located at positive values of the first dimension and were split into two sub groups. HCA yielded three groups, clearly segmenting the samples according to their type of closure, but not according to their turbidity level. The samples C1 bottle1 and C3 bottle1 were without any taint, and were characterized by the terms "woody", "white fruits", "yellow fruits" and "honey". The other group (C1 bottle2, C3 bottle2, C2 bottle1 and C2 bottle2) was characterized by some reduction terms as "dust/cardboard" or "rotten egg/cabbage" as well as "toasted", "citrus" and "butter". The separation between the two bottles of the samples C1 and C3 suggest a bottle effect or maybe also a lack of repeatability of the panel for these two samples.

Contrary to the CA of the 2009 samples (data not given), the CA of the 2010 samples showed a better discrimination between samples and a better consistency between REDOX odor scores and general attributes.

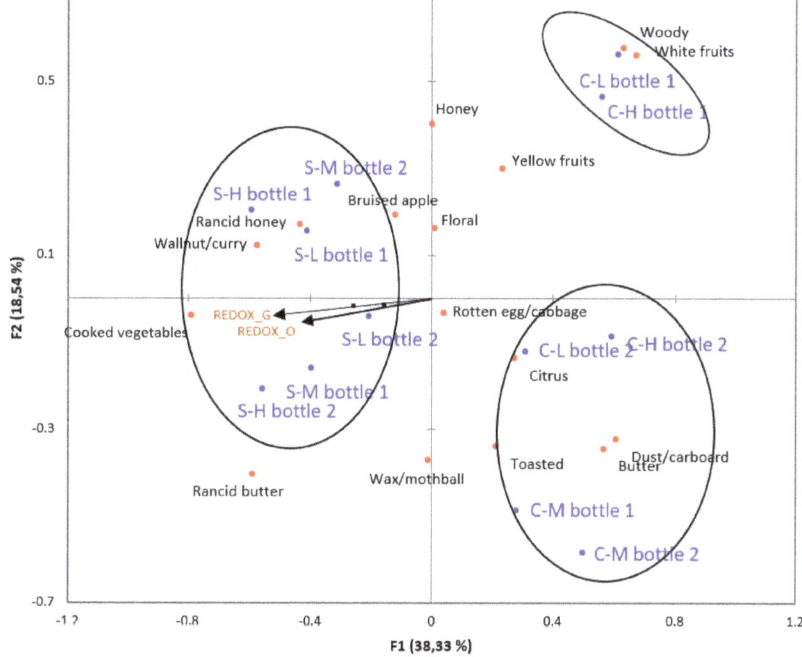

Figure 3. Correspondence Analysis of the general description of the two bottles of each 2010 sample. Ellipses indicate the clusters obtained by HCA using Ward's criterion. O: orthonasal; G: global.

3.2. Comparative Sensory Description of Synthetic Coextruded Stopper Samples

The strong oxidation differences observed between closures on the results of the monadic sensory description are likely to hide the subtle differences between turbidity levels (because of a contrast sensory effect). In order to better explore subtle differences due to turbidity levels, a comparative sensory assessment was carried out only on the three turbidity levels for the synthetic coextruded stopper samples.

The mean values for orthonasal and global REDOX odor scores are presented in Table 5. The results showed that there was a turbidity effect for 2010 samples in orthonasal ($F = 5.34$, $p = 0.011$) and in global ($F = 10.04$, $p = 0.001$) conditions, but no significant effect was found for 2009 samples ($F = 0.845$, $p = 0.44$ for orthonasal and $F = 0.942$, $p = 0.40$ for global conditions).

Table 5. Average scores for orthonasal and global REDOX odor assessment of the three turbidity levels for 2009 and 2010 vintages followed by Newman-Keuls' tests ($\alpha = 5\%$).

Sample Code	2009 Vintage		2010 Vintage	
	REDOX Orthonasal	**REDOX Global**	**REDOX Orthonasal**	**REDOX Global**
S-H	1.46 a	1.01 a	2.69 a	2.57 a
S-M	1.52 a	1.02 a	2.04 ab	2.17 a
S-L	1.93 a	1.48 a	1.47 b	0.93 b

Samples associated with the same letter were not significantly different.

Table 6. Results of the ANOVAS performed on the concentration of polyphenols for 2009 and 2010 samples. Probabilities in bold are significant at α = 5%. F: Fisher F-ratio; $Pr > F$: probability associated to the F-ratio.

Source of Variation	Phenolic Acids		Cinnamic Acids		Flavan-3-ols		GRP		Tyrosol		Total Polyphenols		TPI	
	F	$Pr > F$	F	$Pr > F$	F	$Pr > F$	F	$Pr > F$	F	$Pr > F$	F	$Pr > F$	F	$Pr > F$
vintage	224.9	<0.0001	3711.5	<0.0001	161.89	<0.0001	1334.9	<0.0001	14.05	0.002	3471.11	<0.0001	796.39	<0.0001
closure	3.3	0.09	15.7	0.001	26.05	0.0001	1.63	0.22	1.10	0.31	20.21	0.001	0.36	0.56
turbidity	32.8	<0.0001	0.16	0.85	0.36	0.70	116.6	<0.0001	309.26	<0.0001	35.44	<0.0001	6.68	0.009
vintage*closure	6.5	0.023	0.10	0.76	3.11	0.01	17.94	0.001	14.07	0.002	0.26	0.62	0.59	0.46
vintage*turbidity	2.0	0.17	30.0	<0.0001	0.36	0.706	7.30	0.007	66.77	<0.0001	7.94	0.005	6.38	0.011
closure*turbidity	1.0	0.38	0.28	0.76	0.76	0.484	3.29	0.067	0.468	0.64	0.15	0.87	1.27	0.31

Table 7. Results of the Newman-Keuls' test (α = 5%) performed on the concentration of polyphenols for 2009 and 2010 samples.

Main Effects	Phenolic Acids	Cinnamic Acids	Flavan-3-ols	GRP	Tyrosol	Total Polyphenols	TPI
vintage	Average (mg·L^{-1})	Average (mg·L^{-1})	Average (mg·L^{-1})	Average (mg·L^{-1})	Average (mg·L^{-1})	Average (mg·L^{-1})	Average (mg·L^{-1})
2009	4.54 a	57.47 a	1.30 a	3.56 a	23.49 a	90.37 a	8.39 a
2010	3.93 b	43.09 b	0.29 b	2.88 b	23.156 b	73.36 b	7.29 b
closure	Average (mg·L^{-1})	Average (mg·L^{-1})	Average (mg·L^{-1})	Average (mg·L^{-1})	Average (mg·L^{-1})	Average (mg·L^{-1})	Average (mg·L^{-1})
C	4.27 a	50.75 a	1.00 a	3.21 a	23.28 a	82.5 a	7.85 a
S	4.20 a	49.82 b	0.60 b	3.24 a	23.37 a	81.2 b	7.83 a
turbidity	Average (mg·L^{-1})	Average (mg·L^{-1})	Average (mg·L^{-1})	Average (mg·L^{-1})	Average (mg·L^{-1})	Average (mg·L^{-1})	Average (mg·L^{-1})
L	4.13 b	50.28 a	0.84 a	3.35 a	24.81 a	83.41 a	7.93 a
M	4.47 a	50.21 a	0.77 a	3.29 b	23.01 b	81.75 b	7.84 ab
H	4.11 b	50.37 a	0.78 a	3.03 c	22.15 c	80.44 c	7.75 b

Samples associated with the same letter were not significantly different.

Table 5 shows that for 2010 samples and for both conditions, S-H was perceived as significantly more oxidized than S-L but not than S-M. As for the monadic profile, orthonasal and global REDOX odor scores were significantly correlated ($r = 0.95$, $p < 0.05$).

As for the general description, the results suggest that 2010 samples presented wider oxidation differences than 2009.

3.3. Phenolic Composition

The results of the ANOVA carried out on the phenolic compositions from both 2009 and 2010 vintages are presented in Table 6. Concerning the main effects, vintage reached significance for each class of polyphenols (phenolic acids, cinnamic acids, flavan-3-ols, tyrosol and GRP) and for total polyphenols and TPI. Table 7 shows the post-hoc Newman-Keuls' mean comparison of the average concentrations of all polyphenolic families for each on the main ANOVA factors tested. Table 7 shows that 2009 samples presented significantly higher concentrations for all the families of polyphenols after several years of bottle aging.

Screw caps showed higher concentrations for cinnamic acid, total polyphenols and Flavo-3-ols. However, the latter showed a significant interaction with vintage. Cinnamic acids and total polyphenols showed significant interactions with vintage. On the other hand, phenolic acids and GRP, the major polyphenols in chardonnay wines, were fairly similar between wines under cork or screw cap.

Concerning the effect of turbidity, significance was reached for the concentrations of phenolic acids, GRP, tyrosol, total polyphenols and TPI, but only phenolic acids showed an effect independent of the vintage, with the medium level of turbidity being slightly more concentrated than high and low levels.

3.4. Relationships between Sensory Characteristics and Phenolic Composition

We tried to predict the REDOX scores for orthonasal and global conditions for every vintage from their phenolic concentrations by means of multiple linear regressions. Only variables showing a correlation with the corresponding REDOX scores higher than 0.3 were considered in the linear regressions. The results of the stepwise linear regressions for both vintages and both conditions are presented in Table 8.

Table 8. Results of the multiple regression analysis carried out to predict REDOX odor character by phenolic concentrations. R^2: determination coefficient of the regression; t = Student value; $Pr > |t|$: probability associated with the absolute Student value. – indicates that the variable was not selected for the model because of an $r < 0.3$; ns indicates that the contribution of the variable to the predictive model was not significant. ns*: not significant.

Variable	REDOX 2009 Orthonasal $R^2 = 0.26$; $p = 0.086$		REDOX 2009 Global $R^2 = 0.61$; $p = 0.014$		REDOX 2010 Orthonasal $R^2 = 0.94$; $p < 0.0001$		REDOX 2010 Global $R^2 = 0.92$; $p < 0.0001$	
	t	$Pr > |t|$	t	$Pr > |t|$	t	$Pr > |t|$	t	$Pr > |t|$
Phenol Acids	ns*	ns	ns	ns	–	–	–	–
Cinnamic acids	–	–	ns	ns	ns	ns	–	–
Flavan-3-ols	ns	ns	−3.12	0.012	−10.93	<0.0001	−8.06	<0.0001
GRP	–	–	2.26	0.05	4.12	0.003	6.72	<0.0001
Tyrosol	–	–	ns	ns	–	–	–	–
Total polyphenols	–	–	ns	ns	ns	ns	–	–
TPI	ns	ns	–	–	–	–	–	–

Table 8 shows that no significant predictive model could be obtained from the 2009 orthonasal data. However, the predictive linear model for 2009 global was significant, and the ones for 2010 were very significant. The three significant predictive models were:

$$REDOX\ 2009\text{-}global = -3.54 - 0.93 \times Flavan\text{-}3\text{-}ols + 1.56 \times GRP$$
$$REDOX\ 2010\text{-}orthonasal = -1.932 - 4.08 \times Flavan\text{-}3\text{-}ols + 1.43 \times GRP$$
$$REDOX\ 2010\text{-}global = -3.63 - 2.35 \times Flavan\text{-}3\text{-}ols + 1.82 \times GRP$$

All three predictive models had basically the same pattern, with Flavan-3-ols as negative predictors and GRP as positive predictors. These results suggest that the presence of Flavan-3-ol can predict lower sensory oxidations, whereas high GRP levels predict higher oxidative notes. These results are particularly consistent between vintages for global perception.

4. Discussion

We implemented a combination of sensory and physicochemical analytical strategies in order to better understand the relationship between the wine polyphenol composition and the occurrence of oxidative notes. First of all, it is interesting to note that orthonasal and global assessment conditions were significantly correlated. This result is in agreement with previous findings [13]. However, as seen in Section 3.4, global evaluation enabled better regression models than orthonasal evaluations. One possible reason is that the in-mouth conditions in terms of temperature and contact surface with air favored the release of aroma compounds that were subsequently better perceived retronasally.

According to previous literature, strong vintage and closure effects were expected. Concerning turbidity, no particular hypothesis was formulated. The effects of the studied factors on the chemical and sensory characteristics of the samples are discussed in the following sections.

4.1. Effect of Vintage, Closure and Turbidity on the Wines Polyphenolic Content

The results showed a clear effect of vintage on the concentrations of all of the phenolic classes. Previous studies have already pointed out that for each vintage, a unique polyphenolic composition is conferred upon the resulting wine [26,28]. Flavan-3-ols, catechin and epicatechin were greatly impacted by closure compared to phenolic acids and GRP, likely due to their higher susceptibility to participation in slow oxidative processes during bottle storage [4]. Recent results have consistently shown that an unprecedented diversity of compounds could actually be involved in the discrimination of wine composition according to the type of closure [28]. Finally, must clarification level showed a significant effect on several polyphenolic classes, albeit largely modulated by vintage, but the mechanisms involved are still unclear.

4.2. Vintage Effect on REDOX Perception

Since the two vintages tested were quite different in terms of polyphenolic composition, vintages were analyzed separately for the sensory data. Taken together, the results confirm the importance of the composition of raw material, which is mostly vintage-dependent. Indeed, 2009 samples seemed more resistant to oxidation than those from 2010, while the latter showed wider sensory differences.

4.3. Closure Effect on REDOX Perception

The monadic evaluation clearly showed significant differences between closure types for 2010 samples, but not for 2009. It has been previously shown in the literature that wines closed with synthetic cork are more prone to oxidation than wines closed with screw caps [14,16]. The choice of the stoppers was made in order to increase the probability of having REDOX odor differences between samples. Concerning the general description of the 2009 samples, no segmentation by closure was shown. The 2010 samples showed a clear segmentation by closure type, with the samples closed with screw cap described as fruitier, more oaky and even showing a slight reductive aroma. On the other hand, the samples closed with synthetic cork showed clear signs of oxidation ("rancid honey", "cooked vegetables", "walnut/curry" and "bruised apple"). These results are consistent with previous research [13,14].

4.4. Turbidity Effect on REDOX Perception

The monadic evaluation of 2009 samples showed significant differences in turbidity for both conditions; unfortunately, significant turbidity*wine closure interactions prevent us from interpreting the main effects.

However, concerning the 2010 samples, the panel failed to show a significant turbidity effect, probably because the important differences between types of closures masked more subtle differences between turbidity levels.

A modification in the sensory protocol was implemented in a subsequent sensory session to confirm the existence of significant REDOX odor differences between turbidity levels for both vintages. As a result, a comparative assessment of the synthetic closures showed significant differences only for 2010 samples. Taken together, these results suggest for the first time that must turbidity could have an impact on the oxidative stability of white wines. To date, there are virtually no references concerning the relationship between must clarification and the polyphenol composition of the resulting wine, so the mechanism of this effect is still unclear, and further research is needed. One possible explanation is that, as seen in Section 4.1, must clarification somehow changes the polyphenolic composition of the wine, which could impact its sensory REDOX characteristics.

4.5. Effect of the Phenolic Composition on REDOX Perception

Multiple linear regressions showed significant predictive models of REDOX perception scores for 2010 data (both conditions) and for global perception of 2009 data. For all three models, our results suggest that the presence of Flavan-3-ol could predict lower sensory oxidations, whereas high GRP levels predicts higher oxidative notes. Such results provide an unprecedented model of the oxidation sensory note of dry chardonnay wine based on the concentration of two major polyphenol-related compounds, which is highly consistent with the literature. Flavan-3-ols and GRP have indeed already been described as being involved in oxidation reactions in wine, but never subjected to a sensory analysis. GRP is a well-known polyphenolic compound that appears preferentially during prefermentative steps, where non-protected musts lead to higher concentration of GRP due to the nucleophilic attack of the quinone of caftaric acid with the reduced form of glutathione [29]. Additionally, some studies have also shown that GRP continues to accumulate in wine during bottle aging under oxidative environments [30,31] without knowing exactly its origin. GRP is thus considered a final marker of an historical event of oxidation in must and/or wine. Similarly, the oxidation and browning of white musts have been shown to be largely correlated to the content of hydroxycinnamic acids and favored by the presence of Flavan-3-ols [32,33]. Additionally, Flavan-3-ols are slowly oxidized in wine in presence of oxygen and metal catalysts as clearly described by Danilewicz [5]. Therefore, our results shed new and interesting light on the impact that prefermentative management of musts can have on the later REDOX sensory note of bottle-aged dry chardonnay wines. In particular, they suggest that actions that reduce the production of GRP and maintain high levels of flavanols, such as must protection before alcoholic fermentation, could lead to lower oxidation notes.

5. Conclusions

The present study showed that the sensory oxidative character of bottle-aged chardonnay wines results from complex interactions between the different variables included in the experiment. It also showed the difficulty of generalizing the effects to different vintages. The polyphenolic composition was also quite vintage-dependent, and could partially explain the sensory differences between vintages. The main result of this study is that concentrations of Flavan-3-ols and GRP could both be significant predictors of wines oxidative sensory character, but with opposite effects.

Altogether, these results bring valuable new clues for questioning the impact of oenological practices—from prefermentative steps to bottling—on the subsequent development of oxidation, through the combined analysis of bottle aged wines.

Supplementary Materials: The following are available online at www.mdpi.com/2306-5710/4/1/19/s1. Table S1: raw data corresponding to the sensory sessions; Table S2: raw data corresponding to the polyphenols chemical analysis.

Acknowledgments: The research was funded by the IABECA research federation working on Agriculture, Biodiversity, Environment, Behavior and Food. We thank the Bureau Interprofessionnel des Vins de Bourgogne for the white wine samples. We finally thank the first-year enology students (2015/2016), who kindly participated in our sensory sessions for this experiment.

Author Contributions: Jordi Ballester: conceived and designed the sensory experiments, main writing contributor, sensory data discussion and interpretation. Mathilde Magne: trained the panel and collected the sensory data. Perrine Julien: analyzed the sensory data. Laurence Noret: performed polyphenol analyses. Maria Nikolantonaki: significant contribution in the introduction section. Christian Coelho: conceived and designed the chemical analyses, project leader. Régis D. Gougeon: significant contribution in the interpretation of the results, general supervisor and coordinator.

Conflicts of Interest: The authors declare no conflict of interest.

References

1. Roullier-Gall, C.; Witting, M.; Moritz, F.; Gil, R.B.; Goffette, D.; Valade, M.; Schmitt-Kopplin, P.; Gougeon, R.G.D. Natural oxygenation of champagne wine during ageing on lees: A metabolomics picture of hormesis. *Food Chem.* **2016**, *203*, 207–215. [CrossRef] [PubMed]

2. Lopes, P.; Silva, M.A.; Pons, A.; Tominaga, T.; Lavigne, V.; Saucier, C.; Darriet, P.; Teissedre, P.-L.; Dubourdieu, D. Impact of oxygen dissolved at bottling and transmitted through closures on the composition and sensory properties of a sauvignon blanc wine during bottle storage. *J. Agric. Food Chem.* **2009**, *57*, 10261–10270. [CrossRef] [PubMed]

3. Ugliano, M. Oxygen contribution to wine aroma evolution during bottle aging. *J. Agric. Food Chem.* **2013**, *61*, 6125–6136. [CrossRef] [PubMed]

4. Pons, A.; Nikolantonaki, M.; Lavigne, V.; Shinoda, K.; Dubourdieu, D.; Darriet, P. New insights into intrinsic and extrinsic factors triggering premature aging in white wines. In *Advances in Wine Research*; American Chemical Society: Washington, DC, USA, 2015; Volume 1203, pp. 229–251.

5. Danilewicz, J.C. Interaction of sulfur dioxide, polyphenols, and oxygen in a wine-model system: Central role of iron and copper. *Am. J. Enol. Vitic.* **2007**, *58*, 53.

6. Bueno, M.; Culleré, L.; Cacho, J.; Ferreira, V. Chemical and sensory characterization of oxidative behavior in different wines. *Food Res. Int.* **2010**, *43*, 1423–1428. [CrossRef]

7. Silva Ferreira, A.C.S.; Guedes de Pinho, P.; Rodrigues, P.; Hogg, T. Kinetics of oxidative degradation of white wines and how they are affected by selected technological parameters. *J. Agric. Food Chem.* **2002**, *50*, 5919–5924. [CrossRef]

8. Culleré, L.; Cacho, J.; Ferreira, V. An assessment of the role played by some oxidation-related aldehydes in wine aroma. *J. Agric. Food Chem.* **2007**, *55*, 876–881. [CrossRef] [PubMed]

9. Pripis-Nicolau, L.; de Revel, G.; Bertrand, A.; Maujean, A. Formation of flavor components by the reaction of amino acid and carbonyl compounds in mild conditions. *J. Agric. Food Chem.* **2000**, *48*, 3761–3766. [CrossRef] [PubMed]

10. Rizzi, G.P. Formation of strecker aldehydes from polyphenol derived quinones and α-amino acids in a nonenzymic model system. *J. Agric. Food Chem.* **2006**, *54*, 1893–1897. [CrossRef] [PubMed]

11. Nikolantonaki, M.; Waterhouse, A.L. A method to quantify quinone reaction rates with wine relevant nucleophiles: A key to the understanding of oxidative loss of varietal thiols. *J. Agric. Food Chem.* **2012**, *60*, 8484–8491. [CrossRef] [PubMed]

12. Nikolantonaki, M.; Darriet, P. Identification of ethyl 2-sulfanylacetate as an important off-odor compound in white wines. *J. Agric. Food Chem.* **2011**, *59*, 10191–10199. [CrossRef] [PubMed]

13. Godden, P.; Francis, L.; Field, J.; Gishen, M.; Coulter, A.; Valente, P.; HØJ, P.; Robinson, E. Wine bottle closures: Physical characteristics and effect on composition and sensory properties of a semillon wine 1. Performance up to 20 months post-bottling. *Aust. J. Grape Wine Res.* **2001**, *7*, 64–105. [CrossRef]

14. Skouroumounis, G.; Kwiatkowski, M.; Francis, I.L.; Oakey, H.; Capone, D.L.; Duncan, B.; Sefton, M.A.; Waters, E.J. The impact of closure type and storage conditions on the composition, colour and flavour

properties of a riesling and a wooded chardonnay wine during five years' storage. *Aust. J. Grape Wine Res.* **2005**, *11*, 369–377. [CrossRef]

15. Skouroumounis, G.K.; Kwiatkowski, M.J.; Francis, I.L.; Oakey, H.; Capone, D.L.; Peng, Z.; Duncan, B.; Sefton, M.A.; Waters, E.J. The influence of ascorbic acid on the composition, colour and flavour properties of a riesling and a wooded chardonnay wine during five years' storage. *Aust. J. Grape Wine Res.* **2005**, *11*, 355–368. [CrossRef]

16. Silva, M.A. *Closure Effect on Sensory Quality of Wine*; Université Bordeaux 2: Bordeaux, France, 2011.

17. Guaita, M.; Petrozziello, M.; Motta, S.; Bonello, F.; Cravero, M.C.; Marulli, C.; Bosso, A. Effect of the closure type on the evolution of the physical-chemical and sensory characteristics of a montepulciano d'abruzzo rosé wine. *J. Food Sci.* **2013**, *78*, C160–C169. [CrossRef] [PubMed]

18. Rodrigues, H.; Ballester, J.; Valentin, D. Ageing effect on minerality perception of Chablis wines. In Proceedings of the 10th International Symposium of Enology of Bordeaux, Bordeaux, France, 29 June–1 July 2015; Vigne et vin Publications Internationales: Bordeaux, France; pp. 681–685.

19. Brajkovich, M.; Tibbits, N.; Peron, G.; Lund, C.M.; Dykes, S.I.; Kilmartin, P.A.; Nicolau, L. Effect of screwcap and cork closures on so2 levels and aromas in a sauvignon blanc wine. *J. Agric. Food Chem.* **2005**, *53*, 10006–10011. [CrossRef] [PubMed]

20. Psarra, E.; Makris, D.P.; Kallithraka, S.; Kefalas, P. Evaluation of the antiradical and reducing properties of selected greek white wines: Correlation with polyphenolic composition. *J. Sci. Food Agric.* **2002**, *82*, 1014–1020. [CrossRef]

21. Villaño, D.; Fernández-Pachón, M.S.; Troncoso, A.M.; García-Parrilla, M.C. Influence of enological practices on the antioxidant activity of wines. *Food Chem.* **2006**, *95*, 394–404. [CrossRef]

22. Lachman, J.; Miloslav, Š.; Katerina, F.; Vladimír, P. Major factors influencing antioxidant contents and antioxidant activity in grapes and wines. *Int. J. Wine Res.* **2009**, *1*, 101–121. [CrossRef]

23. Sioumis, N.; Kallithraka, S.; Makris, D.P.; Kefalas, P. Kinetics of browning onset in white wines: Influence of principal redox-active polyphenols and impact on the reducing capacity. *Food Chem.* **2006**, *94*, 98. [CrossRef]

24. Karbowiak, T.; Gougeon, R.D.; Alinc, J.-B.; Brachais, L.; Debeaufort, F.; Voilley, A.; Chassagne, D. Wine oxidation and the role of cork. *Crit. Rev. Food Sci. Nutr.* **2009**, *50*, 20–52. [CrossRef]

25. Campo, E.; Ballester, J.; Langlois, J.; Dacremont, C.; Valentin, D. Comparison of conventional descriptive analysis and a citation frequency-based descriptive method for odor profiling: An application to burgundy pinot noir wines. *Food Qual. Preference* **2010**, *21*, 44–55. [CrossRef]

26. Roullier-Gall, C.; Lucio, M.; Noret, L.; Schmitt-Kopplin, P.; Gougeon, R.D. How subtle is the "terroir" effect? Chemistry-related signatures of two "climats de bourgogne". *PLoS ONE* **2014**, *9*, e97615. [CrossRef] [PubMed]

27. Carando, S.; Teissedre, P.L.; Cabanis, J.C. Hplc coupled with fluorescence detection for the determination of procyanidins in white wines. *Chromatographia* **1999**, *50*, 253–254. [CrossRef]

28. Roullier-Gall, C.; Hemmler, D.; Gonsior, M.; Li, Y.; Nikolantonaki, M.; Aron, A.; Coelho, C.; Gougeon, R.D.; Schmitt-Kopplin, P. Sulfites and the wine metabolome. *Food Chem.* **2017**, *237*, 106–113. [CrossRef] [PubMed]

29. Kritzinger, E.C.; Bauer, F.F.; du Toit, W.J. Role of glutathione in winemaking: A review. *J. Agric. Food Chem.* **2013**, *61*, 269–277. [CrossRef] [PubMed]

30. Vallverdú-Queralt, A.; Verbaere, A.; Meudec, E.; Cheynier, V.; Sommerer, N. Straightforward method to quantify gsh, gssg, grp, and hydroxycinnamic acids in wines by uplc-mrm-ms. *J. Agric. Food Chem.* **2015**, *63*, 142–149. [CrossRef] [PubMed]

31. Coetzee, C.; Van Wyngaard, E.; Šuklje, K.; Silva Ferreira, A.C.; du Toit, W.J. Chemical and sensory study on the evolution of aromatic and nonaromatic compounds during the progressive oxidative storage of a sauvignon blanc wine. *J. Agric. Food Chem.* **2016**, *64*, 7979–7993. [CrossRef] [PubMed]

32. Cheynier, V.; Basire, N.; Rigaud, J. Mechanism of trans-caffeoyltartaric acid and catechin oxidation in model solutions containing grape polyphenoloxidase. *J. Agric. Food Chem.* **1989**, *37*, 1069. [CrossRef]

33. Oszmianski, J.; Cheynier, V.; Moutounet, M. Iron-catalyzed oxidation of (+)-catechin in model systems. *J. Agric. Food Chem.* **1996**, *44*, 1712–1715. [CrossRef]

Article

Application of a Pivot Profile Variant Using CATA Questions in the Development of a Whey-Based Fermented Beverage

Marcelo Miraballes *, Natalia Hodos and Adriana Gámbaro

Área Evaluación Sensorial, Departamento de Ciencia y Tecnología de Alimentos, Facultad de Química, Universidad de la República, General Flores 2124, Montevideo CP 11800, Uruguay; nhodos@fq.edu.uy (N.H.); agambaro@fq.edu.uy (A.G.)
* Correspondence: mmiraballes@fq.edu.uy; Tel.: +598-292-457-35

Received: 15 November 2017; Accepted: 9 January 2018; Published: 1 February 2018

Abstract: During the development of a food product, the application of rapid descriptive sensory methodologies is very useful to determine the influence of different variables on the sensory characteristics of the product under development. The Pivot profile (PP) and a variant of the technique that includes check-all-that-apply questions (PP + CATA) were used for the development of a milk drink fermented from demineralised sweet whey. Starting from a base formula of partially demineralised sweet whey and gelatin, nine samples were elaborated, to which various concentrations of commercial sucrose, modified cassava starch, and whole milk powder were added. Differences in sucrose content affected the sample texture and flavour and the modified starch was able to decrease the fluidity and increase the texture of creaminess and firmness, of the samples. The two applied sensory methodologies achieved good discrimination between the samples and very similar results, although the data analysis was clearly simplified in relation to the difficulty and time consumed in the PP + CATA variant.

Keywords: whey; pivot profile; CATA; fermented beverage

---

1. Introduction

One of the main problems in environmental management of the small and medium dairy industry is the destination of by-products generated in the industrial activities, such as milk whey. Thus, in recent decades, there has been an increase in interest in the use of this by-product [1]. The high content in lactose makes milk whey a raw material (from the industrial viewpoint) with significant potential for the development of fermented products [2]. This application has the advantage that the production process is very similar to the production of, for example, a conventional yoghurt, so the start-up and production cost is not high for a dairy company. The use of lactic acid bacteria in the fermentation of whey is associated with intense bacterial metabolic activity with respect to the carbohydrates, lipids, proteins, and allergenic peptides present in it. Thus, bacteria promote digestibility and preservation. In addition, their action increases the content of lactic acid and other metabolites, such as aromatic compounds that contribute to the flavour, texture, and sweetness of the final product [3]. The sweet whey in its pure form presents a low sensory acceptability, which is due to the unpleasant flavour caused by the high content of mineral salts. Although the fermentation process considerably improves the sensory profile and acceptability of the product, it is not sufficient to achieve acceptability values comparable to those obtained with milk drinks made with milk [4]. In this sense, it is also interesting to use the previously demineralised sweet whey as an input for fermentation, to obtain a product with greater acceptability. To solve this drawback, some authors have produced fermented beverages from sweet whey with the addition of milk at different levels, significantly improving its aroma and flavour characteristics, thus causing an increase in the acceptability of the products [5–8]. In these articles, fermented beverages were made with a different degree of milk substitution with sweet whey. The main limitation found was that, sensorially, the substitution of milk with sweet whey is only viable up to an average of 50% since the acceptability of the product falls afterwards because the texture of the product is not adequate, and unwanted flavour is detected. Regarding the nutritional aspects, fermented milk is an important source of nutrients and provides beneficial health effects, such as stimulation of the immune system, cholesterol reduction, appetite regulation, and a decreased risk of some types of cancer [9,10]. In addition, the organic acids generated during the fermentation process, such as lactic acid, help the absorption of iron from other foods [11].

The sensory profile in the development of a food product is traditionally obtained using a descriptive analysis through a panel of trained assessors [12]. This methodology provides very accurate and reproducible results; however, it has the disadvantage of consuming significant amounts of time and is relatively costly [13]. In this context, several rapid sensory characterisation methodologies have been developed in recent years. These have the advantage of reducing the time needed to obtain results and lower associated costs, in addition to being able to be performed by individuals without prior training [14]. Pivot profile (PP) is a rapid descriptive methodology developed by Thuillier [15] to obtain descriptive information based on the free description technique and is very commonly used in that area. The methodology has been used in the sensory characterisation of champagnes [16] and dairy products, such as chocolate ice cream [17] and Greek yoghurts [18]. Although this methodology allows a very good description of the products evaluated, the data analysis is often very difficult and slow because all the text generated by the assessors must be analysed and interpreted [19]. Check-all-that-apply (CATA) questions are a quick and practical descriptive method to obtain information about how the characteristics of the products are perceived. These questions consist of a list from which individuals choose words or phrases that they consider appropriate for the product they are evaluating [20]. This methodology has been widely used in sensory analysis due to its simplicity and because it is easy, intuitive, and requires less cognitive effort from the participants compared to other techniques [14,21].

In this context, the objective of this work was to compare the results of the PP technique applications and a variant of the PP technique using CATA questions (PP + CATA) with panels of semi-trained assessors in the development of a fermented milk drink from demineralised sweet whey. In this way, we sought to automize the data analysis and avoid the generated text analysis stage, since the terms used to describe the samples in this variant of the technique would be selected in advance and all the assessors would use the same vocabulary.

2. Materials and Methods

2.1. Formulations

For the processing of the samples as a base formula, the following were used: 8% partially demineralised (40% mineral content reduced) and dehydrated sweet whey (Conaprole, Montevideo, Uruguay) and 0.3% gelatin (Boom 220, Bloom, Leiner Davis, NY, USA). Nine samples were developed varying the concentrations of commercial sucrose, modified cassava starch (SuperCorp 75, Horizonte Amidos, PR, Brazil), and whole milk powder (Conaprole, Uruguay), following a factorial design of three variables and two levels (Table 1). In each formulation, drinking water treated through reverse osmosis was used to make up the solutions to 100%. The concentration ranges were selected based on preliminary trials. Sucrose concentrations were chosen to reflect those usually found in commercial fermented beverages, and reported articles in which milk was added to the whey as a method for increasing the acceptability of fermented whey products [8,22–24]. In this way, samples with different sensory characteristics were obtained, both in texture and flavour.

Table 1. Concentrations used in the formulation of samples [1].

Samples	Milk Powder (%)	Starch (%)	Sugar (%)
A	2.5	0	4
B	2.5	0	6
C	2.5	1	4
D	2.5	1	6
E	5.0	0	4
P	5.0	0	6
G	5.0	1	4
H	5.0	1	6

[1] Each formulation contained 8% partially demineralised and dehydrated sweet whey and 0.3% gelatin.

In the present work, sample P was randomly selected (Table 1) as the Pivot. According to [16], the selection of the Pivot sample generates a minimal effect on the results obtained using this methodology and therefore is not a key aspect for the good performance of the method. This criterion was used in a sensory characterisation study of ice cream, in which the Pivot sample was selected at random [17].

2.2. Elaboration Process

The water required for each formulation was heated at 50 °C and then mixed with the solid ingredients for 5 min at 100 rpm. Next, the heat treatment was carried out, bringing the mixture to a temperature of 85 °C, which was maintained for 5 min, stirring at 200 rpm. The mixture was then placed in 1000 mL Durham glass bottles with sterile lids and cooled in a water bath until reaching a temperature of 42 °C. Then, the mixture was inoculated with a lactic ferment dispersion (Yo-Mix 205 LYO 250 DCU, Danisco, France) to obtain an initial concentration in the mixture of 0.2 direct culture units/L. It was gently shaken manually with care so as not to add air. Fermentation was carried out in a temperature controlled oven at 42 ± 1° C. The process was monitored by means of pH and ended when a value of 4.5 was reached. The fermentation time depended on the corresponding formula and was approximately 4 to 5 h. Then, the mixture was cooled to 25 °C in a water bath. Once that temperature was reached, each sample was agitated at 100 rpm for 3 min. Subsequently, it was stored at 4 °C until its evaluation, which was carried out 24 h after. Both the heat treatment and the agitation of the mixture was performed using a Taurus My cook kitchen robot (Taurus S.A., Spain).

2.3. Panels of Semi-Trained Assessors

Twenty individuals were selected who had previous experience in participation as sensory assessors of panels of descriptive analysis of different foods. These individuals were divided into two groups according to the previous experience of each individual in sensory evaluation. This was done so that there would be no difference in their sensory evaluation experience between both groups. Both panels evaluated the samples, but followed two different methodologies. One panel evaluated the samples following the PP methodology, while the other used PP + CATA. All evaluations were conducted in duplicate in two different sessions in a standardised sensory evaluation room according to [25]. RedJade software (RedJade, Redwood Shores, CA, USA) was used as an interface for data collection.

2.4. Pivot Profile

Each panellist received samples in pairs (one identified as the Pivot and another coded with three-digit numbers in alternate presentation order). For each pair, they were asked to write down (in the assessor´s own words) the sensory attributes that they thought the sample had in greater and lesser intensity than the pivot sample, focusing on the texture and flavour attributes. Between the evaluation of samples, each panellist had to drink a little water and wait 30 s before continuing with the next one.

2.5. Pivot Profile + CATA

The form of evaluation following this variant of the methodology was similar to that of PP except that the panellists did not have to write the attributes. Each panellist received two lists of 23 sensory attributes of flavour and texture (Table 2). In the first list, the assessor had to select the attributes they thought the sample had in greater intensity than the pivot. Then, in the second list, the panellist had to select the attributes they thought the sample had in lesser intensity than the pivot.

Table 2. Attributes used in the PP + CATA methodology.

Flavour	Texture
Aftertaste	Rough
Butter flavour	Firmness
Cooked	Liquid
Characteristic flavour	Melting speed
Artificial	Mouth coating
Dairy flavour	Sticky
Strange flavour	Soft
Salty	Lumpy
Sour	Filamentous
Bitter	Gummy
Sweet	Gelatinous
	Creamy

The attributes were selected based on a literature review on possible sensory descriptors present in fermented milk drinks and yogurts, both traditional and with added dairy whey [22,23,26]. To evaluate the reproducibility of both panels, the Pivot sample was used as a blind sample and is observed in the results as P' [27].

2.6. Data Analysis

2.6.1. Pivot Profile

The data were analysed according to [16]; all the attributes generated were grouped semantically into two categories: flavour and texture. Then, words with the same meaning (synonyms) were grouped within the same attribute, for example: astringent, rough, and shrivelled, using a dictionary to identify them. Next, a data table was created as follows. For each sensory attribute defined in the previous stage, we quantified the number of times that the attribute was mentioned as being more intense than the pivot (positive frequency) and the times it was mentioned as less intense than the pivot (negative frequency). The negative frequency was subtracted from the positive frequency, resulting in a data table containing positive and negative values. To obtain a contingency table containing only positive values, the lowest value of the table was added to each value of the table. Thus, the lowest value of the table was transformed into absolute zero and all data in the table were converted to a positive value or zero. From this table, a correspondence analysis (CA) was performed.

2.6.2. Pivot Profile + CATA Data Analysis

In this case, it was not necessary to perform the grouping and classification of attributes because they were already established when using the CATA questions. The frequency of selection of each term of the CATA question used to describe the sample was calculated as more intense than the pivot (positive frequency) and as less intense than the pivot (negative frequency). The negative frequency was subtracted from the positive frequency and the table was transformed so that it only contains values greater than or equal to zero. Then, from the obtained table, a CA was conducted.

2.6.3. Comparison of Obtained Sensory Maps

To compare the results obtained using the PP methodology and its variant using CATA questions, the correlation coefficient Rv between the coordinate matrices of the samples was determined in the first two dimensions of the respective correspondence analysis [28]. In addition, the significance of the Rv coefficient was determined using a permutation test [29]. To visually compare the similarity between the sensory configurations of the samples obtained through both methodologies, a multi factorial analysis (MFA) was performed on the coordinates of the samples in the first two dimensions of the corresponding CA. All statistical analyses were performed using XL-Stat 2017 software (Addinsoft, Paris, France).

3. Results

3.1. Pivot Profile

Figure 1 shows the representation of the samples and attributes generated by the assessors in the description of the samples using the PP technique in the first two factors of the CA. In this case, the first two factors explained almost 91% of the variability of the experimental data obtained.

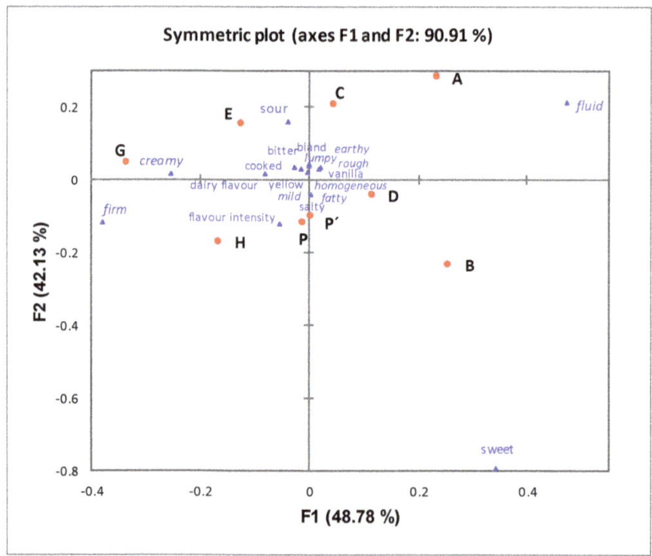

Figure 1. Representation of the samples and attributes in the first two factors of the correspondence analysis of sensory data obtained using Pivot profile (PP). Texture attributes are shown in italics and flavour attributes in regular type.

The attributes that explain the flavour differences found in the samples are the sourness and sweetness. As for the texture, they differed due to their creaminess, firmness, and fluidity.

The first factor, which explained 48.8% of the variability, was associated positively with the fluidity or liquid texture of the samples and negatively regarding creaminess and firmness. These results show coherence in the sensory descriptions obtained, since they are opposing characteristics. On the other hand, the second factor was associated positively with sourness, while negatively with the sweetness of the samples, explaining a variability percentage of 42.1%. As observed in the first factor, the results show relative coherence. As shown in Figure 1, the other attributes are located very close to the origin of the graph coordinates, so these attributes did not serve to discriminate the samples.

Evident differences can be observed in the evaluated samples, since they are dispersed in the four quadrants of the sensory map obtained from the CA (Figure 1). Samples A and B were the most liquid, least creamy, and least firm. In turn, they differed in terms of flavour, with being A the same as C and with E being more sour than B. In addition, sample B was sweeter than the others. These results are consistent with the formulation of the samples shown in Table 1. Samples A and B were prepared with a low level of milk (2.5%) and without modified starch; the only difference was in the sucrose content, which was higher in sample B. Regarding samples C and D, both were characterised as fluid but not as much as for samples A and B. Their difference was also given by the sweetness and sourness, which is consistent because these samples differed in their formulation only in their sucrose content. The difference in the sucrose content used in this study influences the sensory profile of the product. Moreover, the modified starch used reduces fluidity and increases the feeling of creaminess and firmness of the samples. Regarding the texture of samples G and H, these were those of a greater creaminess and firmness and less fluidity. This shows that the samples, when elaborated with a high level of milk (5%) and modified starch, have those characteristics. In turn, sample H was perceived as having a greater sweetness intensity than G, which is consistent with the sucrose content of both samples.

In this case, sample P′ is almost in the same position as the Pivot sample (P). The panel seemingly behaved in a reproducible way.

3.2. Pivot Profile + CATA

The results obtained using the PP + CATA methodology are illustrated in Figure 2. The first two factors explained 89.4% of the variability of the data. The first factor explained 63% of the variability in this case, while the second factor explained 26.3% of the variability.

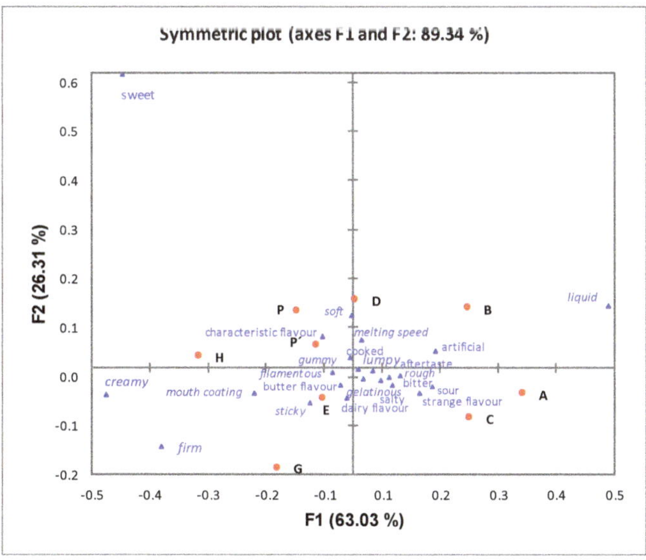

Figure 2. Representation of the samples and attributes in the first two factors of the correspondence analysis of sensory data obtained using PP + CATA. Texture attributes are shown in italics and flavour attributes in regular type.

The attributes that most differentiated the samples in this case were liquid, creamy, firm, and sweet. The first factor was associated positively with the liquid texture attribute and negatively with the creamy and firm attributes, while the second factor was associated positively with sweetness. With respect to the sensory description obtained from the samples, the methodology achieved good discrimination between the samples because they are distributed in the four quadrants of the graph, and their sensory characterization is very similar to that obtained using PP. On the other hand, the positions of samples P and P′ are very close in the map; therefore, the panel behaved in a reproducible way when using this sensory methodology.

3.3. Comparison of Obtained Sensory Maps

The sensory configurations obtained in both sessions are shown in Figure 3. They illustrate that the sensory spaces established through both methodologies are very similar because the samples in both methodologies are located near one another in the plan. On the other hand, a highly significant ($p = 0.001$) value was obtained in the Rv coefficient (0.89) between both configurations.

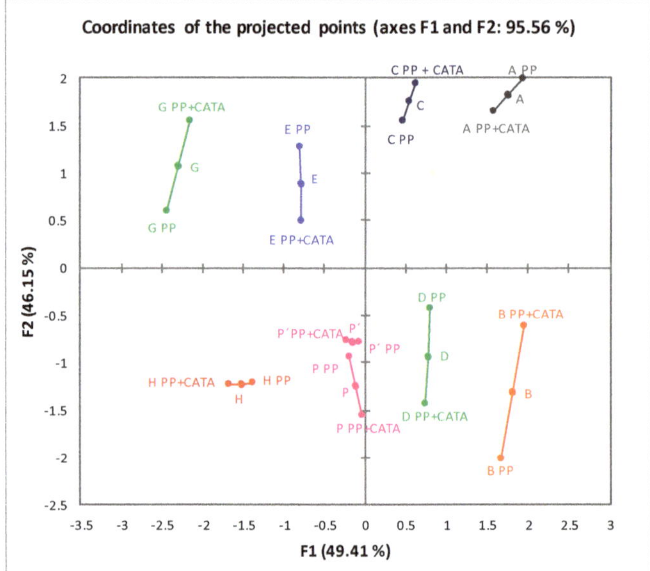

Figure 3. Representation of the superimposed samples in the MFA of both methodologies. All the samples are represented by two points corresponding to each applied methodology (PP and PP + CATA). In turn, a third point, which corresponds to the consensus representation and considers both instances of evaluation, is presented.

4. Discussion

The results obtained through both methodologies were very good in terms of the power of discrimination of the samples. With the PP methodology, it was possible to explain 90.9% of the variability obtained, while this figure was 89.4% when using PP + CATA. At the same time, the attributes by which the samples were discriminated in both methodologies were practically the same. For PP, these were sweetness, sourness, fluidity, creaminess, and consistency. For PP + CATA, the samples were discriminated by these same attributes except for sourness. In this case, the sourness was not very well explained by any of the factors of the CA.

However, in both methodologies, the samples were perceived in a similar way from the sensory point of view and the configurations obtained in both CAs were also similar, which is evident, considering the obtained Rv value herein, 0.89. The Rv coefficient is used as a way of comparing two different factorial configurations. When the value is closer to 1, the correlation these configurations will have is higher, and the closer to 0, the lower it will be. This coefficient depends on the relative position of the points in the configuration and is independent of the rotation and translation [28]. The minimum value to assert that two configurations are relatively similar that has been used in different articles when performing comparisons is between 0.65 and 0.90 and has been widely used in the comparison of data obtained with consumers through PP, CATA questions and Napping [18], PP and free comments [17], Napping in different conditions [30], questions [23,31], different instances of word associations [32], and in performance evaluations of panels of trained sensory assessors [33].

One of the advantages of the use of open response methodologies, such as PP, is the freedom in the language used to describe the samples by the evaluators and the richness in their variety [34]. In this sense, this methodology has an advantage over the variant used in this study using CATA questions (PP + CATA) because, in the latter, it is very important to correctly select the attributes, which is something typical of the CATA question methodology. In contrast, PP has a disadvantage in that the analysis of the generated text to describe the sample stage can be complex and time-consuming, and it has some degree of subjectivity due to researchers' interpretation and categorisation of terms written by assessors [32]. This problem could be solved using the proposed variant (PP + CATA) because the data analysis is clearly simplified, as it is not necessary to perform any text analysis due to the use of the previously predefined language.

The results obtained in this study are similar in terms of the sensory description of the samples and the factorial spaces obtained using semi-trained judges; thus, the proposed PP variant is likely to be useful for the evaluation of dairy products, such as those used in this study.

5. Conclusions

The application of a rapid descriptive sensory methodology, such as PP, enables the evaluation of the influence of different ingredients and concentrations of them in the sensory characteristics of the samples. With the variant used, which combines PP with CATA questions, the results were very similar to those obtained when using PP, but the data analysis was clearly simplified by eliminating the text analysis. Notwithstanding, future studies that compare the results obtained when using PP and the proposed variant (PP + CATA) in other products that compare the results of both techniques obtained with semi-trained assessors and with consumers should be conducted.

Author Contributions: M.M. and A.G. conceived and designed the experiments; M.M. and N.H. performed the experiments; M.M. and N.H. analyzed the data; M.M. and A.G. wrote the paper.

Conflicts of Interest: The authors declare no conflict of interest.

References

1. Ganju, S.; Gogate, P.R. A review on approaches for efficient recovery of whey proteins from dairy industry effluents. *J. Food Eng.* **2017**, *215*, 84–96. [CrossRef]
2. Sanmartín, B.; Díaz, O.; Rodríguez-Turienzo, L.; Cobos, A. Composition of caprine whey protein concentrates produced by membrane technology after clarification of cheese whey. *Small Rumin. Res.* **2012**, *105*, 186–192. [CrossRef]
3. Mauriello, G.; Moio, L.; Moschetti, G.; Piombino, P.; Addeo, F.; Coppola, S. Characterization of lactic acid bacteria strains on the basis of neutral volatile compounds produced in whey. *J. Appl. Microbiol.* **2001**, *90*, 928–942. [CrossRef] [PubMed]
4. Soares, D.S.; Fai, A.E.; Oliveira, A.M.; Pires, E.M.; Stamford, T.L. Aproveitamento de soro de queijo para produção de iogurte probiótico. *Arq. Bras. Med. Vet. Zootec.* **2011**, *63*, 996–1002. [CrossRef]

5. Caldeira, L.A.; Ferrão, S.P.B.; Fernandes, S.A.D.A.; Magnavita, A.P.A.; Santos, T.D.R. Desenvolvimento de bebida láctea sabor morango utilizando diferentes níveis de iogurte e soro lácteo obtidos com leite de búfala. *Ciência Rural* **2010**, *40*, 2193–2198. [CrossRef]

6. Moreira, R.W.M.; Madrona, G.S.; Branco, I.G.; Bergamasco, R.; Pereira, N.C. Avaliação sensorial e reológica de uma bebida achocolatada elaborada a partir de extrato hidrossolúvel de soja e soro de queijo. *Acta Sci. Technol.* **2010**, *32*, 435–438.

7. Oliveira, G.I.C.; Resende, L.M.; Soares, E.M.; Matos, S.P. Alimentação e suplementação de ferro em uma população de lactentes carentes. *Pediatria (São Paulo)* **2006**, *28*, 18–25.

8. Rakin, M.; Bulatovic, M.; Zaric, D.; Stamenkovic-Djokovic, M.; Krunic, T.; Boric, M.; Vukasinovic-Sekulic, M. Quality of fermented whey beverage with milk. *Hem. Ind.* **2016**, *70*, 91–98. [CrossRef]

9. Haug, A.; Høstmark, A.T.; Harstad, O.M. Bovine milk in human nutrition—A review. *Lipids Health Dis.* **2007**, *6*, 25. [CrossRef] [PubMed]

10. Wolf, I.V.; Vénica, C.I.; Perotti, M.C. Effect of reduction of lactose in yogurts by addition of β-galactosidase enzyme on volatile compound profile and quality parameters. *Int. J. Food Sci. Technol.* **2015**, *50*, 1076–1082. [CrossRef]

11. Branca, F.; Rossi, L. The role of fermented milk in complementary feeding of young children: Lessons from transition countries. *Eur. J. Clin. Nutr.* **2002**, *56*, S16–S20. [CrossRef] [PubMed]

12. Delarue, J.J.; Lawlor, B.; Rogeaux, M. *Rapid Sensory Profiling Techniques and Related Methods, Applications in New Product Development and Consumer Research*; Woodhead Publishing: Sawston, UK, 2015; ISBN 9781845694760.

13. Varela, P.; Ares, G. *Novel Techniques in Sensory Characterization and Consumer Profiling*; CRC Press: Boca Raton, FL, USA, 2014; ISBN 978-1-4665-6630-9.

14. Varela, P.; Ares, G. Sensory profiling, the blurred line between sensory and consumer science. *A review of novel methods for product characterization. Food Res. Int.* **2012**, *48*, 893–908. [CrossRef]

15. Thuillier, B. *Rôle du CO₂ dans L'appréciation Organoleptique des Champagnes—Expérimentation et Apports Méthodologiques*; Université de Reims: Reims, France, 2007.

16. Thuillier, B.; Valentin, D.; Marchal, R.; Dacremont, C. Pivot© profile: A new descriptive method based on free description. *Food Qual. Prefer.* **2015**, *42*, 66–77. [CrossRef]

17. Fonseca, F.G.A.; Esmerino, E.A.; Filho, E.R.T.; Ferraz, J.P.; da Cruz, A.G.; Bolini, H.M.A. Novel and successful free comments method for sensory characterization of chocolate ice cream: A comparative study between pivot profile and comment analysis. *J. Dairy Sci.* **2016**, *99*, 3408–3420. [CrossRef] [PubMed]

18. Esmerino, E.A.; Tavares Filho, E.R.; Thomas Carr, B.; Ferraz, J.P.; Silva, H.L.A.; Pinto, L.P.F.; Freitas, M.Q.; Cruz, A.G.; Bolini, H.M.A. Consumer-based product characterization using Pivot Profile, Projective Mapping and Check-all-that-apply (CATA): A comparative case with Greek yogurt samples. *Food Res. Int.* **2017**, *99*, 375–384. [CrossRef] [PubMed]

19. Valentin, D.; Chollet, S.; Lelie, M.; Lelièvre, M.; Abdi, H. Quick and dirty but still pretty good: A review of new descriptive methods in food science. *Int. J. Food Sci. Technol.* **2012**, *47*, 1563–1578. [CrossRef]

20. Ares, G.; Jaeger, S.R. Check-all-that-apply (CATA) questions with consumers in practice: Experimental considerations and impact on outcome. In *Rapid Sensory Profiling Techniques*; Elsevier: Amsterdam, The Netherlands, 2015; pp. 227–245, ISBN 9781782422488.

21. Dos Santos, B.A.; Bastianello Campagnol, P.C.; da Cruz, A.G.; Galvão, M.T.E.L.; Monteiro, R.A.; Wagner, R.; Pollonio, M.A.R. Check all that apply and free listing to describe the sensory characteristics of low sodium dry fermented sausages: Comparison with trained panel. *Food Res. Int.* **2015**, *76*, 725–734. [CrossRef] [PubMed]

22. Bruzzone, F.; Ares, G.; Giménez, A. Temporal aspects of yoghurt texture perception. *Int. Dairy J.* **2013**, *29*, 124–134. [CrossRef]

23. Cadena, R.S.; Caimi, D.; Jaunarena, I.; Lorenzo, I.; Vidal, L.; Ares, G.; Deliza, R.; Giménez, A. Comparison of rapid sensory characterization methodologies for the development of functional yogurts. *Food Res. Int.* **2014**, *64*, 446–455. [CrossRef]

24. Zoellner, S.S.; Cruz, A.G.; Faria, J.A.F.; Bolini, H.M.A.; Moura, M.R.L.; Carvalho, L.M.J.; Sant'ana, A.S. Whey beverage with acai pulp as a food carrier of probiotic bacteria. *Aust. J. Dairy Technol.* **2009**, *64*, 177.

25. International Organization Standardization. *Sensory Analysis: General Guidance for the Design of Test Rooms*; International Organization Standardization: Geneva, Switzerland, 1988.

26. Gallardo-Escamilla, F.J.; Kelly, A.L.; Delahunty, C.M. Influence of starter culture on flavor and headspace volatile profiles of fermented whey and whey produced from fermented milk. *J. Dairy Sci.* **2005**, *88*, 3745–3753. [CrossRef]

27. Liu, J.; Grønbeck, M.S.; Di Monaco, R.; Giacalone, D.; Bredie, W.L.P. Performance of Flash Profile and Napping with and without training for describing small sensory differences in a model wine. *Food Qual. Prefer.* **2016**, *48*, 41–49. [CrossRef]

28. Robert, P.; Escoufier, Y. A unifying tool for linear multivariate statistical methods: The RV-coefficient. *Appl. Stat.* **1976**, *25*, 257–265. [CrossRef]

29. Josse, J.; Pagès, J.; Husson, F. Testing the significance of the RV coefficient. *Comput. Stat. Data Anal.* **2008**, *53*, 82–91. [CrossRef]

30. Miraballes, M.; Fiszman, S.; Gámbaro, A.; Varela, P. Consumer perceptions of satiating and meal replacement bars, built up from cues in packaging information, health claims and nutritional claims. *Food Res. Int.* **2014**, *64*, 456–464. [CrossRef]

31. Bruzzone, F.; Ares, G.; Giménez, A. Consumers' texture perception of milk desserts. Comparison with trained assessors' data. *J. Texture Stud.* **2012**, *43*, 214–226. [CrossRef]

32. Pontual, I.; Amaral, G.V.; Esmerino, E.A.; Pimentel, T.C.; Freitas, M.Q.; Fukuda, R.K.; Sant'Ana, I.L.; Silva, L.G.; Cruz, A.G. Assessing consumer expectations about pizza: A study on celiac and non-celiac individuals using the word association technique. *Food Res. Int.* **2017**, *94*, 1–5. [CrossRef] [PubMed]

33. Blancher, G.; Clavier, B.; Egoroff, C.; Duineveld, K.; Parcon, J. A method to investigate the stability of a sorting map. *Food Qual. Prefer.* **2012**, *23*, 36–43. [CrossRef]

34. Ten Kleij, F.; Musters, P.A. Text analysis of open-ended survey responses: A complementary method to preference mapping. *Food Qual. Prefer.* **2003**, *14*, 43–52. [CrossRef]

Article

Effect of Different Glass Shapes and Size on the Time Course of Dissolved Oxygen in Wines during Simulated Tasting

Parpinello Giuseppina Paola [1], Meglioli Matteo [2], Ricci Arianna [1] and Versari Andrea [1,*

[1] Department of Agricultural and Food Sciences, University of Bologna, Piazza Goidanich 60, 47521 Cesena (FC), Italy; giusi.parpinello@unibo.it (P.G.P.); arianna.ricci4@unibo.it (R.A.)
[2] Mosti Mondiale Inc., 6865 Route 132, Sainte-Catherine, QC J5C 1B6, Canada; matteo.meglioli@mostimondiale.com
* Correspondence: andrea.versari@unibo.it; Tel.: +39-0547-338-124

Academic Editor: Manuel Malfeito Ferreira
Received: 21 November 2017; Accepted: 20 December 2017; Published: 4 January 2018

Abstract: The different shapes and sizes of wine glass are claimed to balance the different wine aromas in the headspace, enhancing the olfactory perception and providing an adequate level of oxygenation. Although the measurement of dissolved oxygen in winemaking has recently received much focus, the role of oxygen in wine tasting needs to be further disclosed. This preliminary study aims to explore, for the first time, the effect of swirling glasses of different shapes and sizes on the oxygen content of wine. Experimental trials were designed to simulate real wine tasting conditions. The O_2 content after glass swirling was affected to a considerable extent by both the type of wine and the glass shape. A lack of correlation between the shape parameters of five glasses and the O_2 content in wine was found which suggests that the nonequilibrium condition can occur during wine tasting. The International Standard Organisation (ISO) glass—considered to be optimal for the wine tasting—allowed less wine oxygenation than any other glass shapes; and the apparent superiority of the ISO glass is tentatively attributed to the more stable oxygen content with time; i.e., less variability in oxygen content than any other glass shape.

Keywords: glass swirling; glass shape; nonequilibrium conditions; oxygen sensor; wine tasting

---

1. Introduction

Although the critical role of oxygen in enology has recently been disclosed from a chemistry [1] and winemaking point of view [2–4], little information is available from the sensory perspective.

Before tasting, the glass of wine is usually "swirled" by holding the glass by the stem and gently rotating it. This action, technically called 'orbital shaking', increases the surface area of the wine by spreading it over the inner part of the glass and consequently enabling some evaporation to take place [5]. Moreover, it is also expected to draw in some oxygen from the air. The ingress estimated on the undisturbed surface of a wine is about 200 mg/h/m^2 [6].

The physics of wine swirling was recently investigated with an elegant fluid dynamic approach, which modelled the pumping mechanism induced by the wave propagation along the glass wall [7]. Three factors seemed to determine whether the team spotted one big wave in the wine or several smaller ripples: (i) the ratio of the level of wine poured in to the diameter of the glass; (ii) the ratio of the diameter of the glass to the width of the circular shaking; (iii) and the ratio of the forces acting on the wine. From a practical point of view, these findings suggest that the mixing and oxygenation may be optimized with an appropriate choice of shaking diameter (d) and rotation speed (rpm). In this view, the glass shape parameters (Figure S1) can play a key role as they may influence the perceived volume of wine [8], and the perception of wine odors [9–11], and color [12–14], and therefore the consumer's preference [15,16] as well. Moreover, with time, the glass shape affects the change of headspace chemical composition of wine poured inside, and the D ratio (i.e., maximum diameter divided by opening diameter) seems to be the most important parameter relating glass shape to headspace composition [17].

Considering that both consumers and professional wine tasters usually swirl the glass of wine for approx. 10 to 20 s to unlock odors, there is a need for information to disclose the effect of glass shape on the oxygen content of wines during simulated tasting condition.

This preliminary trial aimed to study whether the glass shape can affect the oxygen content in wine under both static and dynamic conditions (i.e., swirling), the latter to simulate the standard procedure of sensory evaluation of wine. The use of optical oxygen sensors (also called minisensors) allowed for the first time the on-line non-invasive and non-destructive oxygen measurements in a glass under dynamic conditions.

2. Materials and Methods

2.1. Samples

Both red and white wines were selected for this study, including (i) a Rebola 'Nita' white wine Colli di Rimini DOP 2013, (ii) a Sangiovese red wine Carlo Leo Romagna DOP 2013, and (iii) a Cabernet Sauvignon 'Tano' Colli di Rimini DOP 2013 (Az. Agric. Le Calastre, Rimini, Italy). Sangiovese is the main red grape in Italy, Cabernet Sauvignon is a well know international grape variety, and Rebola is an emerging local white grape variety of great interest as well. The bottled wines were provided by the producer and were stored at room temperature (20 ± 1 °C) until the swirling trials. Preliminary characterization of wine composition (Table S1) was carried out according to endorsed methods (RESOLUTION OIV/OENO 390/2010).

2.2. Oxygen Measurement

Non-invasive dissolved oxygen (DO) in wine was measured using an OXY-4 oxygen meter (PreSens GmbH, Regensburg, Germany) equipped with a polymer optical fiber and PSt3 spot, also called minisensor (Presens GmbH). The PSt3 spot had a thickness of 1 mm, a diameter of 4 mm, and a response time (t_{90}, the time for 90% of the change in signal to occur) of 10 s. In each glass, one minisensor was glued 5 mm below the wine level and calibrated with water at a controlled temperature according to the manufacturer's instructions (Figure S2). In each bottle, once uncorked, the dissolved O_2 content of wine was directly measured with an oxygen-dipping probe (Presens GmbH) placed in the middle of the bottle in a dynamic regime (i.e., with stirring). Before bottling the wines, three glass bottles (750 mL) were equipped with two minisensors each to ascertain both the headspace and the dissolved O_2 content in static regime (i.e., bottled wines).

2.3. Glasses

Six different wine glasses were selected for this study, including the ISO official-tasting glass (Bormioli, Parma, Italy) (Table 1; Figure 1). Glass parameters were measured according to the literature (Hirson, Heymann and Ebeler 2012) and each glass was filled with a fixed volume of wine (50 mL) to fit the conditions commonly used during professional wine tasting [18].

2.4. Swirling Trials

To simulate the gesture of hand swirling during wine tasting, each glass of wine was placed onto an orbital shaker (model 709/R ASAL, Cernusco s/N, Italy) with a shaking diameter of 3 cm and a shaking speed of 150 rpm (this value was optimized with preliminary screening trails from 100 to 250 rpm). Each glass of wine was swirled for 10, 20 and 40 s, using independent wine samples.

Table 1. Glass shape parameters, sample volumes and mass transfer coefficient (k_{La}) for dissolved oxygen trials.

No.	Glass Name	Opening Diameter (mm)	Maximum Diameter (mm)	Wine Diameter (mm)	Height (mm)	Headspace Height (mm)	D Ratio *	Fill (mL)	Headspace Volume (mL)	k_{La} (s⁻¹)
1	ISO	45.2	63.9	63.5	94.5	68.1	1.41	50	165	0.0001
2	Fresh	49.7	79.0	71.7	116.1	89.2	1.59	50	330	0.0003
3	Mature	53.8	83.2	73.9	128.3	100.7	1.55	50	440	0.0002
4	Rich	67.6	89.8	78.2	130.6	105.7	1.33	50	540	0.0005
5	Reserve	73.0	95.8	81.6	142.5	115.8	1.31	50	710	0.0005
6	Super800	73.4	116.1	89.6	107.7	85.8	1.58	50	750	0.0003

* Maximum diameter divided by opening diameter.

Figure 1. Glasses tested for the experimental trials (see Table 1 for details).

2.5. Experimental Design and Statistical Analysis

The four variables (6 glass types, 3 wines, 3-time readings, 3 replicates for each glass) required 162 independent measurements. Statistical analysis was based on a paired *t*-test (*p*-level < 0.05) using the Unscrambler X (v. 10.3, Camo ASA, Oslo, Norway).

3. Results and Discussion

The oxygen content was measured (i) before and after opening the bottle of wine and (ii) before and during the glass swirling trails, for which protocol was designed to simulate the usual wine tasting by consumers and experts. As expected, the concentration of oxygen in bottled wine was very low with an average range from 4 to 14 µg/L for headspace O_2, and in the range of 3 to 29 µg/L for dissolved O_2 in wine. In bottled wine, the rate of O_2 dissolution was less than the consumption; however, once the bottles are opened the wine comes into contact with air and the oxygen content is expected to rise. In fact, soon after opening the O_2 content in bottled wine increased regularly up to 0.99 mg/L in 15 min (Figure 2). These findings are consistent with the initial oxygen absorption capacity of Madiran red wines with pH 3.78 [19]. The O_2 accumulation in wine implied that the rate of oxygen dissolution was higher than the rate of its uptake. The latter can be (indirectly) measured by the drop in SO_2. According to Boulton [20], the oxygen consumption reactions in wine involve the phenolic compounds as the main substrates and the oxygen consumption is the rate-limiting reaction. The rates of this reaction are first order in oxygen concentration and catalyzed by ferrous ion; therefore, the rate constant would be related to the ferrous ion concentration, but the rate law would depend only on the oxygen concentration. The amount of oxygen found after 15 min most likely increases the wine redox value of ca. 25 mV and will theoretically consume ca. 4 mg/L of sulfur dioxide [21]; both parameters could affect the sensory properties of wines to some extent [4,13].

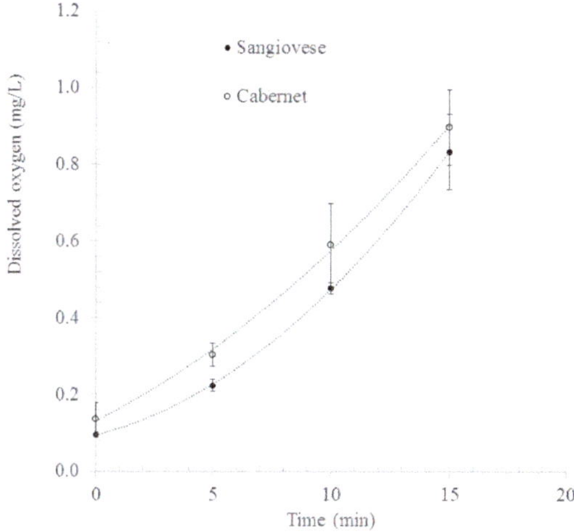

Figure 2. Time course of dissolved oxygen in wine after opening (time zero).

As time progresses, the dissolved oxygen content in wine is expected to reach a plateau approaching the saturation value of ca. 8.6 mg/L at 20 °C [22]. This postulate was confirmed by monitoring with time the dissolved O_2 in a wine glass under firm conditions, i.e., unstirred, unshaken, directly after pouring (Figure 3). The glass n. 5 showed a very fast O_2 intake followed by the glass n. 4, whereas the wine poured on the ISO glass (No. 1), commonly used in professional wine tasting, showed the lowest and most stable O_2 content. Although dissolved oxygen mostly depends on the surface area of wine exposed and the exposure time, a lack of significant correlation was found between the glass shape parameters and the O_2 content in wine. To explain this result, the hypotheses of nonequilibrium conditions and complex interaction among parameters were formulated.

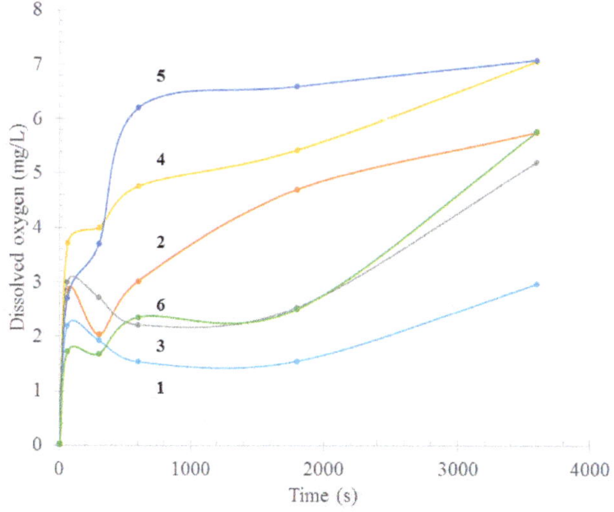

Figure 3. Time course of oxygen dissolution in wine glass under firm conditions. Legend: lines 1–6 refer to different glass shapes (see Table 1 for details).

For any specific wine, the variation of the dissolved oxygen concentration with time can be arranged as:

$$\frac{dO_2}{dt} = k_L a \cdot (O_2^* - O_2)$$

where: $k_L a$: volumetric mass transfer coefficient (T^{-1}); O_2 Dissolved oxygen concentration in wine (mg O_2 L^{-1}); O_2^* Oxygen equilibrium concentration in wine (mg O_2 L^{-1}).

Therefore, the volumetric mass transfer coefficient is the aggregate result of both contributions: the resistance to mass transport in the liquid side (k_L) and the interfacial area (a).

The oxygen transfer rate will decrease during the period of aeration as O_2 approaches O_2^* due to the decline in the driving force ($O_2^* - O_2$). The experimental k_{La} values of wines in opened bottle and glasses were tentatively estimated upon monitoring the increase in the dissolved oxygen concentration of a wine during aeration and agitation and by plotting of oxygen deficit ($O_2^* - O_2$) versus time (t) on a semilogarithmic graph. The slope of the regression line determined the overall mass transfer coefficient ($k_L a$), which showed 5 times variation in the range 0.0001–0.0005 s^{-1} (Table 1), with k_{La} of wine in open bottle at the low value of 0.0001 s^{-1}. As the estimated k_{La} values refer to static conditions (i.e., non-agitated vessels—bottle or glass—and lack of gas supply, e.g., micro-oxygenation) the current findings are consistent with data from the literature. The mass transfer coefficient (k_{La}) in the aerated model wine solution range between 0.00145 s^{-1} [23] and 0.013 s^{-1} [24], the values of which are affected by several parameters, including the oxygen gas flow rate and the intensity of agitation. The best estimate of the rotational speeds effect is given in terms of liquid side mass transfer coefficient (k_L) which increases more than 9 times as rotational speed increases from 50 to 120 rpm [25]. Remarkably, k_{La} is found to increase following a power law with exponent rising from 0.5 to 3 due to significant increases in the interfacial area due to vortex [26].

During professional wine tasting for appellation certification, the wine is held in a glass for about 5–10 min and swirled for approx. 5–30 s before tasting. In this view, the swirling trails were designed to verify the likely occurrence of nonequilibrium conditions during the sensory evaluation of wine. The initial O_2 content at time zero (T_0) was always measured before each swirling trials for every glass shape. The O_2 content after glass swirling was affected to a considerable extent by both the type of wine and the glass shape. Short swirling time (up to 20 s) most often decreased the dissolved oxygen in glass wine (Table 2), while O_2 significantly increased at T_{40} in Rebola white and Sangiovese red wines for all glass shape, except for glass n. 1 in Sangiovese trial. In contrast, the O_2 content in Cabernet Sauvignon wine increased only in glass n. 5 and n. 6 at T_{40}. It seems that the Cabernet Sauvignon would require more time to enhance the O_2 content compared to the other two wines, which are lower in polyphenolic compounds and SO_2.

According to the Henry's law, the oxygen uptake resulting from the presence of antioxidants in wine is rapid and follows a largely exponential form [20]. The alternative hypothesis that O_2 is initially stripped out from solution by the release of CO_2 is postulated. Clearly, the ISO glass (No. 1)—usually considered to be optimal for wine tasting—allowed less wine oxygenation than any other glass shape. Considering that the O_2 content of wine most likely affects the performance of sensory evaluation, based on our findings, the apparent superiority of the ISO glass is tentatively attributed to the more stable oxygen content with time, i.e., less variable than any other glass shape.

The current results under the dynamic regime substantiate the findings of Venturi et al. [27] who also investigated the influence of glass shape on the dissolved oxygen content of a rosé wine under static conditions—using the polarographic ADI dO$_2$ sensor—for which equilibration time was found to be approx. 1 h.

In conclusion, the current preliminary study showed that pouring wine into a glass considerably affects the O_2 content and the likely occurrence of nonequilibrium condition requires a careful standardization procedure during a real wine tasting.

Table 2. Heatmap plot of dissolved oxygen (mg/L) in wine glasses at time zero (T_0) and after 10, 20 and 40 s (T_{10}, T_{20}, T_{40}). Legend: glass 1–6 refers to different shape (see Table 1 for details). The symbol in letters (a, b) refers to significant paired difference between samples (p-level = 0.05) if occurred.

Rebola White Wine: Dissolved O_2 (mg/L)					
Glass	T_0	T_{10}	T_0	T_{20}	T_0	T_{40}
No. 1	3.0	2.9	2.9	3.0	2.7	3.1
No. 2	3.1 [a]	2.2 [b]	4.3	4.1	3.4 [a]	4.2 [b]
No. 3	4.2	4.2	3.1 [a]	4.3 [b]	3.1 [a]	4.5 [b]
No. 4	3.0	3.0	3.1	3.2	3.8 [a]	4.7 [b]
No. 5	3.9 [a]	3.0 [b]	3.7	3.5	3.5 [a]	4.2 [b]
No. 6	2.8	2.5	3.0	3.3	4.5 [a]	6.0 [b]

Sangiovese Red Wine: Dissolved O_2 (mg/L)					
Glass	T_0	T_{10}	T_0	T_{20}	T_0	T_{40}
No. 1	2.8	2.8	2.9 [a]	2.3 [b]	3.5	3.1
No. 2	3.2	3.4	3.2	2.9	2.7 [a]	3.6 [b]
No. 3	2.8	3.3	3.3	3.0	3.3 [a]	4.0 [b]
No. 4	3.1	2.7	3.3	2.9	3.0 [a]	4.2 [b]
No. 5	3.4	2.8	3.2 [a]	4.2 [b]	3.1 [a]	4.2 [b]
No. 6	2.8	3.0	3.1	3.1	3.8 [a]	5.0 [b]

Cabernet Sauvignon Red Wine: Dissolved O_2 (mg/L)					
Glass	T_0	T_{10}	T_0	T_{20}	T_0	T_{40}
No. 1	2.8	2.8	2.9	2.3	2.9	3.0
No. 2	4.2 [a]	3.0 [b]	3.9 [a]	2.8 [b]	4.2 [a]	3.3 [a]
No. 3	3.0	2.7	4.0 [a]	3.1 [b]	4.1	3.8
No. 4	3.5	3.0	3.2	2.8	4.6	4.1
No. 5	3.2	3.1	3.8	3.4	3.8 [a]	4.5 [a]
No. 6	3.4	3.5	3.4	3.8	4.3 [a]	5.5 [a]

Supplementary Materials: The following are available online at www.mdpi.com/2306-5710/4/1/3/s1. Figure S1: Glass parameters, Table S1: Chemical composition of wines, Figure S2: Experimental online measurement.

Acknowledgments: Authors thanks Martelli E. (Nomacorc Italia) for technical assistance.

Author Contributions: V.A. and P.G.P. conceived and designed the experiments; R.A. and M.M. performed the experiments; V.A. analyzed the data; M.M. contributed reagents/materials/analysis tools; P.G.P. wrote the paper.

References

1. Danilewicz, J.C. Review of oxidative processes in wine and value of reduction potentials in enology. *Am. J. Enol. Vitic.* **2011**, *63*, 1–10. [CrossRef]

2. Calderón, J.F.; del Alamo-Sanza, M.; Nevares, I.; Laurie, F. The influence of selected winemaking equipment and operations on the concentration of dissolved oxygen in wines. *Cienc. Investig. Agric.* **2014**, *41*, 273–280. [CrossRef]

3. Oliveira, V.; Lopes, P.; Cabral, M.; Pereira, H. Kinetics of oxygen ingress into wine bottles closed with natural cork stoppers of different qualities. *Am. J. Enol. Vitic.* **2013**, *64*, 395–399. [CrossRef]

4. Ugliano, M. Oxygen contribution to wine aroma evolution during bottle aging. *J. Agric. Food Chem.* **2013**, *61*, 6125–6136. [CrossRef] [PubMed]

5. Arakawa, T.; Iitani, K.; Wang, X.; Kajiro, T.; Toma, K.; Yano, K.; Mitsubayashi, K. A sniffer-camera for imaging of ethanol vaporization from wine: The effect of wine glass shape. *Analyst* **2015**, *140*, 2881–2886. [CrossRef] [PubMed]

6. Mueller-Spath, H. Production of white table wines with minimal SO_2. In Proceedings of the Second Australian Wine Industry Technical Conference Proceedings, Tanunda, South Australia, 7–9 August 1973; Australian Wine Research Institute: Tanunda, South Australia, 1973; pp. 51–52.

7. Reclari, M.; Dreyer, M.; Tissot, S.; Obreschkow, S.; Wurm, F.M.; Farhat, M. Surface wave dynamics in orbital shaken cylindrical containers. *Phys. Fluids* **2014**, *26*. [CrossRef]

8. Pechey, R.; Attwood, A.S.; Couturier, D.-L.; Munafò, M.R.; Scott-Samuel, N.E.; Woods, A.; Marteau, T.M. Does glass size and shape influence judgements of the volume of wine? *PLoS ONE* **2015**, *10*, e0144536. [CrossRef] [PubMed]

9. Delwiche, J.F.; Pelchat, M.L. Influence of glass shape on wine aroma. *J. Sens. Stud.* **2002**, *17*, 19–28. [CrossRef]

10. Hummel, T.; Delwiche, J.F.; Schmidt, C.; Hüttenbrink, K.B. Effects of the form of glasses on the perception of wine flavors: A study in untrained subjects. *Appetite* **2003**, *41*, 197–202. [CrossRef]

11. Vilanova, M.A.R.; Vidal, P.; Cortes, S. Effect of the glass shape on flavor perception of 'toasted wine' from Ribeiro (NW Spain). *J. Sens. Stud.* **2008**, *23*, 114–124. [CrossRef]

12. Cliff, M.A. Influence of wine glass shape on perceived aroma and colour intensity in wines. *J. Wine Res.* **2001**, *12*, 39–46. [CrossRef]

13. Venturi, F.; Andrich, G.; Sanmartin, C.; Scalabrelli, G.; Ferroni, G.; Zinnai, A. The expression of a full-bodied red wine as a function of the characteristics of the glass utilized for the tasting. *CyTA J. Food* **2014**, *12*, 291–297. [CrossRef]

14. Wan, X.; Woods, A.T.; Seoul, K.-H.; Butcher, N.; Spence, C. When the shape of the glass influences the flavour associated with a coloured beverage: Evidence from consumers in three countries. *Food Qual. Prefer.* **2015**, *39*, 109–116. [CrossRef]

15. Spence, C.; Wan, X. Beverage perception and consumption: The influence of the container on the perception of the contents. *Food Qual. Prefer.* **2015**, *39*, 131–140. [CrossRef]

16. Spence, C.; Van Doorn, G. Does the shape of the drinking receptacle influence taste/flavour perception? A review. *Beverages* **2017**, *3*, 33. [CrossRef]

17. Hirson, G.D.; Heymann, H.; Ebeler, S.E. Equilibration time and glass shape effects on chemical and sensory properties of wine. *Am. J. Enol. Vitic.* **2012**, *63*, 515–521. [CrossRef]

18. Jackson, R.S. *Wine Tasting: A Professional Handbook*, 2nd ed.; Academic Press: San Diego, CA, USA, 2009.

19. Moutounet, M.; Mazauric, J.-P. L'oxygène dissous dans les vins. *Revue Francaise d'Oenologie* **2001**, *186*, 12–15.

20. Boulton, R. Both White Wines and Red Wines Can Consume Oxygen at Similar Rates. 2011. Available online: http://www.acenologia.com/enfoques/roger_boulton_enf0612_eng.htm (accessed on 7 December 2017).

21. Ribéreau-Gayon, P.; Glories, Y.; Maujean, A.; Dubourdieu, D. Aging red wines in vat and barrel: Phenomena occurring during aging. In *Handbook of Enology*; John Wiley & Sons, Ltd.: Hoboken, NJ, USA, 2006; pp. 387–428.

22. Kilmartin, P.A. The oxidation of red and white wines and its impact on wine aroma. *Chem. N. Z.* **2009**, *1*, 18–22.

23. Adoua, R.; Mietton-Peuchot, M.; Milisic, V. Modelling of oxygen transfer in wines. *Chem. Eng. Sci.* **2010**, *65*, 5455–5463. [CrossRef]

24. Saa, P.A.; Moenne, M.I.; Perez-Correa, J.R.; Agosin, E. Modeling oxygen dissolution and biological uptake during pulse oxygen additions in oenological fermentations. *Bioprocess Biosyst. Eng.* **2012**, *35*, 1167–1178. [CrossRef] [PubMed]

25. Hebrard, G.; Zeng, J.; Loubiere, K. Effect of surfactants on liquid side mass transfer coefficients: A new insight. *Chem. Eng. J.* **2009**, *148*, 132–138. [CrossRef]

26. Scargiali, F.; Busciglio, A.; Grisafi, F.; Brucato, A. Oxygen transfer performance of unbaffled stirred vessels in view of their use as biochemical reactors for animal cell growth. *Chem. Eng. Trans.* **2012**, *27*, 205–210.

27. Venturi, F.; Andrich, G.; Sanmartin, C.; Taglieri, I.; Scalabrelli, G.; Ferroni, G.; Zinnai, A. Glass and wine: A good example of the deep relationship between drinkware and beverage. *J. Wine Res.* **2016**, *27*, 153–171. [CrossRef]

Communication

Bittersweet Findings: Round Cups Fail to Induce Sweeter Taste

Casparus J. A. Machiels

Department A & F Marketing—Consumer Psychology, Faculty of Agricultural and Nutritional Sciences, Kiel University, Wilhelm-Seelig-Platz 6/7, 24098 Kiel, Germany; jasper.machiels@ae.uni-kiel.de; Tel.: +49-430-880-5153

Received: 1 November 2017; Accepted: 17 January 2018; Published: 1 February 2018

Abstract: An increasing body of literature demonstrates that consumers associate visual information with specific gustatory elements. This phenomenon is better known as cross-modal correspondence. A specific correspondence that has received attention of late is the one between round forms and sweet taste. Research indicates that roundness (as opposed to angularity) is consistently associated with an increased sweetness perception. Focusing on two different cup forms (round versus angular), two studies tested this association for a butter milk drink and a mate-based soft drink. Results, however, were not able to corroborate the frequently suggested correspondence effect, but a correspondence was found between the angular cup and a more bitter taste for the soft drink. These results are discussed in light of previous findings matching sweetness with roundness and bitterness with angularity, hopefully aiding researchers in this field in conducting future experiments.

Keywords: cross-modal correspondence; shape; taste; beverage

1. Introduction

Recent research in sensory marketing highlights the fact that forms and visuals can influence perception of foods, a phenomenon known as cross-modal correspondence [1]. Specifically, perception through one or more senses, such as vision and touch, can influence perception in other modalities, such as taste [2] (see [3] for a recent review specifically related to beverages). Across a range of food products, round forms are believed to be associated with a sweeter taste [1], which is corroborated through a number of experiments [4]. Although such effects are readily discerned in exploratory research and shape-taste matching studies [4,5], it also seems to hold for empirical research involving actual food products. For instance, sweeter beers, as well as sweeter chocolate, are more strongly associated with round shapes [6,7]. These studies, however, rely mostly on controlled laboratory conditions and within-participants designs, where consumers can carefully deliberate associations between sweetness and roundness across shapes. Recent research, however, extends these laboratory results to actual retail settings by showing that consumers, in fact, perceive hot drinks (coffee and chocolate) consumed from cups with round surface patterns to be sweeter than when consumed from cups with angular surface patterns [8].

Empirical evidence thus indicates the non-randomness of associations between round shapes and sweet tastes. Here, I report two studies concerning container shape effects, designed to corroborate the "round equals increased sweetness" heuristic using two different beverages; a butter milk drink and a mate-based soft drink. Contrasting previous work, both studies failed to find differences in terms of sweetness perception. The taste of the mate-based soft drink, however, was found to be bitterer when consumed from an angular cup. In reporting these results, I aim to aid cross-modal correspondence researchers in making better choices in designing future experiments, and as such advance science's cumulative nature in this important field.

2. Study 1

2.1. Materials and Methods

2.1.1. Design, Participants, Stimuli, and Procedure

The study used a one-factor (drinking cup: angular versus round) between-subjects design. A total of 86 consumers (58.1% female, mean age = 46.89, SD = 19.47, range = 18–79) were intercepted randomly (i.e., every fifth visitor) upon entering a large, local supermarket, and were invited to participate in the tasting of a strawberry-flavored butter milk drink. Strawberry-flavored butter milk was chosen for its general non-conspicuousness, its fruit-flavored sweetness, and with the intent to extend the evidence base of cross-modal correspondences to other product categories. No specific ex- or inclusion criteria were applied, but, before participating, all participants were informed about the product and the voluntariness of their participation, and were fully debriefed afterwards. The drinking cups were hard transparent polystyrene cups, chosen based on their similarity in measurements and volume (see Figure 1 for the two cups). To ensure that participants were unaware of the manipulation, they were randomly assigned to either the round or the angular cup in batches of ten (i.e., first to tenth participant drank from the round, eleventh to twentieth from the angular cup, etcetera). Participants completed the questionnaire one after another using the same bar table situated near the entrance of the supermarket. In both conditions they sampled an identical premium brand of strawberry-flavored butter milk, and care was taken for all samples to include exactly the same amount of drink (40 mL). Temperature of the drink was kept at a constant 7 degrees Celsius.

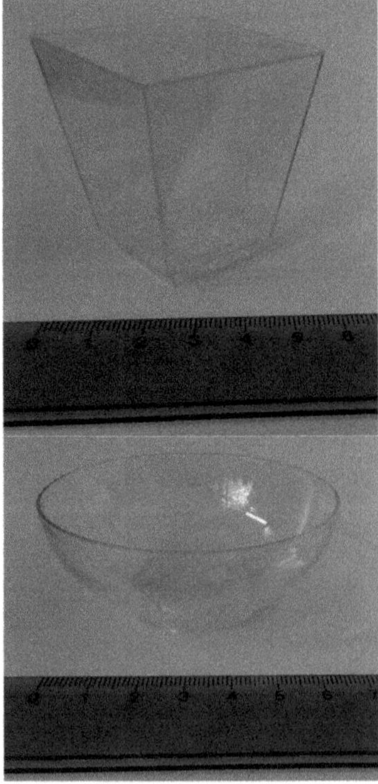

Figure 1. The angular (**above**) and round (**below**) cups used in the Study. Note: ruler measurement is metric (cm).

2.1.2. Measures and Analyses

To ascertain that participants held the cups long enough to get a feeling of the cup's form, they were first asked to answer questions regarding the smell of the product, assessed through the items *strong*, *powerful*, and *intense* (α = 0.83 [8,9]). More importantly, perceived sweetness was assessed through a single item (i.e., *this drink has a sweet taste*). Since sweetened butter milk has both sweet as well as sour taste components, and because angularity can be associated with sourness [10], a related question assessed how sour the drink was perceived (i.e., *this drink has a sour taste*). Complementing basic taste perception, participants additionally assessed the strawberry flavor of the drink with the items *this drink … … tastes fruity, … has a dominant strawberry flavor*, and *… has a distinct strawberry flavor* (α = 0.87). Given that product form angularity is also related to intensity [9], taste intensity was assessed through the items *strong*, *powerful*, and *intense* (α = 0.82 [8,9]). Accounting for subjective differences, general product liking was measured through the item *this product tastes very good*. To control for possible differences in taste experience based on the premise that angular cups are commonly not used as drinking vessels, participants were asked to what extent they thought the drinking experience was *comical* (reverse scored) and *pleasant* (α = 0.76). General preference for butter milk was measured through the items *I generally like butter milk* and *I often drink butter milk* (α = 0.85). The questionnaire ended with questions regarding age and gender. All measures were 7-point Likert scales, ranging from 1 (completely disagree) to 7 (completely agree). Descriptives, frequencies, and between-group differences were calculated, where applicable, through analyses of variance, using IBM SPSS Statistics 22.

2.2. Results

Four consumers did not understand the questionnaire, or submitted non-usable answers (e.g., highlighting whole blocks of question items); their data sets were eliminated, reducing the analysis sample to 82 (57.3% female, mean age in years = 45.91, SD = 19.37, range = 18–79, two participants did not report their age). Preliminary analyses of variance showed that there were no differences between the two groups regarding age and gender (both F's < 1), nor in general liking of butter milk and drinking experience (both F's < 1).

Table 1 lists respondent age and gender distribution, together with means and standard deviations for all dependent variables, per group. Although slightly redundant (given these figures), a multiple analysis of variance, testing differences in the dependent variables depending on the independent variable (cup form: angular versus round) showed no differences between groups for any of the dependent variables (all F's < 1). P- and F-values have been omitted from Table 1 due to the obviousness of non-differences between the group means. Additional—explorative—analyses investigating between-group differences showed no contingence on moderator variables (i.e., gender, butter milk liking, and drinking experience).

Table 1. Age and gender distributions across groups, and means (standard deviations) for all dependent variables (Study 1).

	Group Characteristics				Outcomes					
Manipulation	N (Female)	Age	Sweetness	Sourness	Smell Intensity	Taste Intensity	Liking	Strawberry Flavor	Butter Milk Preference	Drink Experience
Round cup	39 (23)	46.18 (21.58)[1]	4.34 (1.84)[1]	2.72 (1.49)	4.16 (1.48)	4.14 (1.29)[3]	4.05 (1.84)	4.19 (1.60)	3.14 (1.78)	5.24 (1.61)[3]
Angular cup	43 (24)	45.67 (17.40)[1]	4.12 (1.79)	2.72 (1.55)	4.31 (1.54)	4.15 (1.20)	4.33 (1.67)	4.38 (1.39)	3.37 (1.77)	5.03 (1.58)
Overall	82 (47)	45.91 (19.37)[2]	4.22 (1.80)[1]	2.72 (1.51)	4.24 (1.50)	4.14 (1.24)[3]	4.20 (1.75)	4.29 (1.48)	3.26 (1.77)	5.13 (1.58)[3]

Note: superscripts denote number of missing values; Statistical values are based on univariate analyses to account for missing values.

2.3. Discussion

Eighty-two supermarket visitors sampled a strawberry-flavored butter milk drink from either a round or an angular cup. Against expectations, consumer responses showed little sign of the cross-modal correspondence effect between round shapes and sweet taste. One reason for this unexpected finding could lie with the product. Although the mean values for sweetness perception show no indication of this, the approximately 11% sugar content may have been too sweet, limiting variance for effects to occur. Another possibility could trace back to the specific mouth feel associated with butter milk, for instance its creamy consistency. There is evidence that creaminess correlates with sweetness [11,12], which may have influenced participants' sweetness perception. At this point, however, both explanations are speculative only.

Another reason could be that for round shapes to actually change taste perception, more is needed than just the association between round shapes and sweet taste. This means that Study 1's findings are not meaningful with regard to whether consumers actually relied on this specific correspondence. It can only be concluded that no downstream influences on taste perception took place. Wang, Reinoso Carvalho, Persoone, and Spence [11] report on a study where pre-tasting evaluations of round versus angular pieces of chocolate indeed showed a round equals increased sweetness bias, but effects vanished when participants actually tasted the food [11]. As Wang's study manipulated the shape of the actual food, possible priming effects were lost when the food was eaten, indicating the possible necessity of multisensory feedback during consumption (i.e., haptic and visual; [8]). Their findings, together with the basic cross-modal correspondence claim as discussed in the Introduction here, might, then, indicate that this specific correspondence possibly exists on multiple levels. A 'weak' claim, associating round shapes with sweetness, and a 'strong' claim, that sweetness ratings are changed due to exposure to roundness (cf. the different classes of correspondences in [1]). If this assumption were correct, consumers might have associated the round cups with sweetness, but the strength of this association was insufficient to change sweetness perception.

These limitations motivated Study 2, which investigates a different product category, and additionally measures pre-consumption taste expectations in order to incorporate both the weak and strong claim detailed above.

3. Study 2

3.1. Materials and Methods

Similar to Study 1, Study 2 used a one-factor (drinking cup: angular versus round) between-subjects design. This time a convenience student sample was used ($n = 89$), consisting of students of two Introductory Marketing courses (73% female, mean age in years = 23.31, SD = 2.88, range = 18–32). As in Study 1, students were informed beforehand about the product and the voluntariness of their participation, and were fully debriefed afterwards. All students completed the questionnaire simultaneously, while seated in the same auditorium. They tasted a caffeinated soft drink based on mate extract from either a round or an angular cup. The cups were identical to the ones used in Study 1 (see Figure 1). Choice of beverage was based on the premise that the soft drink constitutes another beverage category, is non-creamy, would fit well with the student sample, and is less sweet (appr. 5.7% sugar) than the buttermilk drink sampled in Study 1. Additionally, paralleling sweetness, the drink also incorporates bitter taste notes. The drink is common in Germany, and is comparable to the brands Materva and Club Mate available in the US. Samples were again standardized in volume (40 mL) and temperature (room temperature).

3.2. Measures and Analyses

One limitation of Study 1 was that it did not measure whether participants associated roundness with sweetness, in addition to possible downstream effects on taste perception. In an attempt to overcome this limitation, Study 2 required participants to rate their expectation of the taste of the drink before tasting it [11]. Taste expectation was assessed through the items *I expect this drink to taste . . . . . . sweet, . . . sour,* and *. . . bitter.* Measures regarding the product's smell (*strong, powerful,* and *intense* [8,9], $\alpha = 0.81$), liking (*this product tastes very good*), taste intensity (*strong, powerful,* and *intense* [8,9], $\alpha = 0.89$), mate-based soft drink preference (*I generally like mate drinks* and *I often drink mate drinks,* $\alpha = 0.83$), and drinking experience (*comical* (reverse scored) and *pleasant,* $\alpha = 0.87$) mirrored those of Study 1. It could be argued that the round cups might additionally be associated with smoothness. Therefore, two additional items assessed taste smoothness (i.e., *smooth* and *rough* (reverse scored), $\alpha = 0.70$, note that the German words used in the questionnaire were *mild* and *herb*. The adjective *rough* is used here in the absence of a better alternative.) The taste items (*sweet* and *sour*) were complemented with *bitter*. As in Study 1, all measures were Likert scales, ranging from 1 (completely disagree) to 7 (completely agree), and statistical analyses (i.e., analyses of variance) were calculated using IBM SPSS Statistics 22. As such, Study 2 differed from Study 1 in its sample (students, versus supermarket visitors in Study 1) and beverage (soft drink, versus a buttermilk drink in Study 1).

3.3. Results

Preliminary analyses of variance indicated no differences between the two experimental groups in terms of age ($F(1,83) = 2.18$, ns), gender, general liking of mate-based drinks, and drink experience (all F's < 1).

Table 2 lists group characteristics, along with mean scores and standard deviations for all dependent variables, separately for each group. No differences emerge between the two groups regarding expected and actual sweet taste, or the intensity measures (all F's < 1). However, two findings are noteworthy; the taste of the drink consumed from the round (rather than the angular) cup is perceived as being more smooth (difference of 0.40) and less bitter (difference of 0.76). Analysis of variance shows the difference for *bitter* to be significant at $\alpha = 0.05$ ($F(1,87) = 4.45$, $p = 0.038$, $\eta_p^2 = 0.05$; results for *smooth*: $F(1,87) = 1.33$, $p = 0.252$, $\eta_p^2 = 0.02$).

Table 2. Age and gender distributions across groups, and means (standard deviations) for all dependent variables (Study 2).

Group Characteristics			Outcomes											
Manipulation	N (Female)	Age	Expectations			Actual taste			Smell Intensity	Taste Intensity	Taste Smoothness	Liking	Mate Preference	Drink Experience
			Sweetness	Sourness	Bitterness	Sweetness	Sourness	Bitterness						
Round cup	48 (36) [3]	22.85 (3.20) [2]	5.29 (1.38)	2.92 (1.62)	3.08 (1.82)	4.38 (1.54)	2.54 (1.62)	2.83 (1.67)	4.11 (1.25)	3.75 (1.34)	4.02 (1.67)	4.17 (1.51)	3.14 (1.66) [1]	4.57 (1.81)
Angular cup	41 (29)	23.83 (2.42)	5.07 (1.57)	2.80 (1.32) [1]	3.13 (1.51)	4.46 (1.60)	2.49 (1.72)	3.59 (1.69)	4.06 (1.32)	4.02 (1.32)	3.62 (1.58)	3.93 (1.83)	3.16 (1.99)	4.59 (1.84)
Overall	89 (65) [3]	23.31 (2.88) [2]	5.19 (1.47)	2.86 (1.49) [1]	3.10 (1.67)	4.42 (1.55)	2.52 (1.66)	3.18 (1.71)	4.09 (1.28)	3.87 (1.33)	3.84 (1.63)	4.06 (1.66)	3.15 (1.81) [1]	4.58 (1.82)

Note: superscript denotes number of missing values; Statistical values are based on univariate analyses to account for missing values.

3.4. Discussion

Using a beverage from a different category, the findings from Study 2 shed additional light on possible cross-modal correspondence effects between shapes and tastes [1]. The round cup, as in Study 1, failed to induce increased taste perception of sweetness and visual expectations based only on the cup's shape were also not found. However, the drink in the angular cup was perceived as more bitter in taste, a finding in line with the cross-modal correspondence effect relating angular shapes to bitterness [6,8]. Although the small effect size warrants caution, this positive finding might be interpreted as adding to the growing body of research on this topic. Additionally, although not significant, the difference of .40 for smoothness between the two groups might be interesting for future research to explore in more depth.

4. General Discussion

Correspondences across modalities (i.e., between vision and taste) are increasingly being researched and tested, with reports indicating a seemingly fundamental link between round shapes and sweet taste [1,13]. The goal of the studies reported here was to corroborate these findings by showing that beverages consumed from a cup with round design elements (i.e., a round cup) were perceived as sweeter than an identical beverage consumed from a cup with angular design elements (i.e., an angular cup). However, the results of the two experiments showed little evidence for the correspondence between round shapes and sweet taste. Moreover, not only was the drink in the round cup not perceived as sweeter, the angular cup failed to induce a more intense taste experience [8–10]. Additionally, the well-documented human preference for rounded designs and design features [14,15] did not increase liking for consumers drinking from the round cup. Or, more accurately, if there indeed was an increased liking for one of the cup's shape, this did not spill over to liking for the actual product.

Although these studies were mainly focused on sweetness perception, the results of Study 2 regarding bitterness show that drinking vessel shape has the ability to change basic taste perception. Hence, Study 2's findings, positively equating angularity with a more bitter taste, may readily be interpreted as additional evidence for this specific cross-modal correspondence [6], and the general discernibility of such effects in between-subject experiments (that is, without a direct comparison [8]). Interestingly, these findings are in line with a recent study that was unable to find effects of round latté art shapes on sweetness expectation in coffee, but where the effects of angular shapes on bitterness were found ([16], experiment 3). It is possible that the effects of angular shapes might be more deeply ingrained in the human mind, due to their association with threat [15]. However, assuming that the effect found is actually there, one has to conclude that individual cross-modal effects seem highly susceptible to change depending on the beverage category or sample population. One also has to wonder, then, why expectation effects (based on the beverage's visual inspection) did not surface here [11,16]. It is possible that other influences, like beverage color, might have played a larger role [17]. This limits the practical relevance of using shapes to communicate specific tastes, indicating that front-of-package design elements (i.e., typefaces [18,19] or simple shapes [14]) may not be sufficient to steer consumer response (i.e., expectations of taste) at the point of sale.

This leaves us to account for the null results regarding shape and sweetness across the two beverage categories. The first explanation involves the cups used; although it is to date unclear why cross-modal correspondences exist [13], it cannot be excluded that the round cup was insufficient to induce (the expected) sweetness in taste (and for the angular cup to induce intensity, for that matter [9]). Some might even argue that the cups used were not round enough to engage participants in what leads to cross-modality effects on sweetness perception. Indeed, although the round cups are probably as round as they can get, the haptic properties of the round and angular cups might not have been distinctively different. The 3D-molds used in van Rompay et al.'s study [8], where effects between round shapes and sweetness did show, had angular and round properties along the cups' complete outer surface. Contrast this with the cups used here, where participants might not have experienced the different haptic feedback of angularity and roundness, but instead may have only felt the same smooth surface in both conditions. This might imply that, without the ability to rely on explicit visual contrasts (i.e., other, different shapes), visual perception alone is ineffective in establishing cross-modal correspondences between round shapes and sweet tastes without sufficient congruent haptic information. This might also have been a reason for other null results reported in recent studies investigating the relationship between round shapes and sweet taste in beverages [16,20].

A second explanation might lie within the product category. As mentioned, two recent studies also failed to confirm the hypothesized round equals increased sweetness correspondence [16,20], one for curved (versus straight-edged) beer glasses [20], the other for round (versus angular) shapes of chocolate sprinkled on coffee [16]. Although butter milk (Study 1), mate-based soft drinks (Study 2), coffee ([16], cf. 8) and beer [20] constitute only a fraction of the possible beverage categories, finding null results across these four beverage categories might indicate that these beverages are not suitable for investigating sweetness differences on the basis of visio-haptic information (roundness). Cross-modality effects between round shapes and sweet tastes might, of course, exist perhaps only for food products where taste consists of an intricate interplay of sweet, sour, or bitter components. Positive results, not relying on direct comparisons of shapes, have been reported for coffee and chocolate milk [8], although it is unclear how prominent the role of haptic feedback may have been. Therefore, it cannot be excluded that shapes elicit specific tastes only when consumers are actively engaging in taste searching on the basis of information provided, for example, when consciously examining new products [1], or when specifically instructed to do so in a laboratory setting. Moreover, effects for beverages are bound to rely on product-extrinsic cues like cups or packaging elements (e.g., see [21] for an overview of the different ways beverage containers may influence perception, for instance by changing how the liquid flows through the mouth), whereas for other food categories, like chocolates [11,12], the shape of the product itself can be manipulated.

The third implication of these null results might be the non-replicability of this specific correspondence effect for beverages. Although this possibility does not imply that the general tendency to equate round shapes with sweet taste is non-existent, it does limit the practical usefulness of varying shape properties to change taste perception. Moreover, effects of round shapes on increased sweetness perception have been reported to rely on boundary conditions such as color [17], hedonic preference [4], emotional valence [5], or even on the way answer scales are anchored [22], highlighting the difficulty of finding correspondence effects between shape and taste. As such, it might be a good idea for future research to focus on replication studies—either direct or conceptual—to solidify the foundation of cross-modal correspondence research.

The main limitation of the studies reported here, admittedly, is their sample size. Depending on effect size, in between-subjects experiments, the roughly forty participants per group may not have been enough to discern effects where they exist [23,24]. This limitation, then, also extends to the positive findings regarding bitterness in Study 2, which might either be easier discernable than effects for sweetness and therefore easier to find in smaller samples, or might constitute a statistical artifact.

5. Conclusions

Two studies focused on whether consumers' sweetness perception of two drinks could be altered by the shape of the cup the drinks were consumed from. Both studies failed to generate evidence for the commonly reported correspondence between round shapes and sweet tastes, raising important questions for future cross-modal correspondence research. It still remains unclear how, and to what extent, consumers use shapes as cues for a product's (expected) taste [1].

Readers should note that, on the basis of these two experiments, I am not claiming that this particular cross-modal correspondence effect is nonexistent. I am, however, cautioning against an over-reliance on relative effects (i.e., explicit contrasts) for its practical relevance. As such, it appears important that results like the present ones are made available to aid cross-modal correspondence researchers in further clarifying if, how, and when cross-modal effects are to be expected, notwithstanding making its practical relevance more clear.

Acknowledgments: Special thanks go to Lasse Bahnsen for his excellent research assistance and to Ulrich R. Orth for his helpful comments on previous versions of this manuscript. The author is also grateful for the very helpful suggestions of the anonymous *Beverages* reviewers.

Conflicts of Interest: The author declares no conflict of interest.

References

1. Velasco, C.; Woods, A.T.; Petit, O.; Cheok, A.D.; Spence, C. Crossmodal correspondences between taste and shape, and their implications for product packaging: A review. *Food Qual. Prefer.* **2016**, *52*, 17–26. [CrossRef]
2. Piqueras-Fiszman, B.; Spence, C. Sensory expectations based on product-extrinsic food cues: An interdisciplinary review of the empirical evidence and theoretical accounts. *Food Qual. Prefer.* **2015**, *40*, 165–179. [CrossRef]
3. Spence, C.; van Doorn, G. Does the shape of the drinking receptacle influence taste/flavour perception? A review. *Beverages* **2017**, *3*, 33. [CrossRef]
4. Velasco, C.; Woods, A.T.; Deroy, O.; Spence, C. Hedonic mediation of the crossmodal correspondence between taste and shape. *Food Qual. Prefer.* **2015**, *41*, 151–158. [CrossRef]
5. Salgado-Montejo, A.; Alvarado, J.A.; Velasco, C.; Salgado, C.J.; Hasse, K.; Spence, C. The sweetest thing: The influence of angularity, symmetry, and the number of elements on shape-valence and shape-taste matches. *Front. Psychol.* **2015**, *6*, 1382. [CrossRef] [PubMed]
6. Ngo, M.K.; Misra, R.; Spence, C. Assessing the shapes and speech sounds that people associate with chocolate samples varying in cocoa content. *Food Qual. Prefer.* **2011**, *22*, 567–572. [CrossRef]
7. Deroy, O.; Valentin, D. Tasting liquid shapes: Investigating the sensory basis of cross-modal correspondences. *Chemosens. Percept.* **2011**, *4*, 80–90. [CrossRef]
8. Van Rompay, T.J.L.; Finger, F.; Saakes, D.; Fenko, A. "See me, feel me": Effects of 3D-printed surface patterns on beverage evaluation. *Food Qual. Prefer.* **2017**, *62*, 332–339. [CrossRef]
9. Becker, L.; van Rompay, T.J.L.; Schifferstein, H.N.; Galetzka, M. Tough package, strong taste: The influence of packaging design on taste impressions and product evaluations. *Food Qual. Prefer.* **2011**, *22*, 17–23. [CrossRef]
10. Velasco, C.; Salgado-Montejo, A.; Marmolejo-Ramos, F.; Spence, C. Predictive packaging design: Tasting shapes, typefaces, names, and sounds. *Food Qual. Prefer.* **2014**, *34*, 88–95. [CrossRef]
11. Wang, Q.J.; Reinoso Carvalho, F.; Persoone, D.; Spence, C. Assessing the effect of shape on the evaluation of expected and actual chocolate flavour. *Flavour* **2017**, *6*, 2. [CrossRef]
12. Reinoso Carvalho, F.; Wang, Q.J.; van Ee, R.; Persoone, D.; Spence, C. "Smooth operator": Music modulates the perceived creaminess, sweetness, and bitterness of chocolate. *Appetite* **2017**, *108*, 383–390. [CrossRef] [PubMed]
13. Spence, C.; Ngo, M. Assessing the shape symbolism of the taste, flavour, and texture of foods and beverages. *Flavour* **2012**, *1*, 12. [CrossRef]
14. Westerman, S.J.; Gardner, P.H.; Sutherland, E.J.; White, T.; Jordan, K.; Watts, D.; Wells, S. Product design: Preference for rounded versus angular design elements. *Psychol. Mark.* **2012**, *29*, 595–605. [CrossRef]
15. Bar, M.; Neta, M. Humans prefer curved visual objects. *Psychol. Sci.* **2006**, *17*, 645–648. [CrossRef] [PubMed]
16. Van Doorn, G.; Colonna-Dashwood, M.; Hudd-Baillie, R.; Spence, C. Latté art influences both the expected and rated value of milk-based coffee drinks. *J. Sens. Stud.* **2015**, *30*, 305–315. [CrossRef]
17. Stewart, P.C.; Goss, E. Plate shape and colour interact to influence taste and quality judgments. *Flavour* **2013**, *2*, 27. [CrossRef]
18. Karnal, N.; Machiels, C.J.A.; Orth, U.R.; Mai, R. Healthy by design, but only when in focus: Communicating non-verbal health cues through symbolic meaning in packaging. *Food Qual. Prefer.* **2016**, *52*, 106–119. [CrossRef]
19. Velasco, C.; Woods, A.T.; Hyndman, S.; Spence, C. The taste of typeface. *i-Perception* **2015**, *6*. [CrossRef] [PubMed]
20. Mirabito, A.; Oliphant, M.; van Doorn, G.; Watson, S.; Spence, C. Glass shape influences the flavour of beer. *Food Qual. Prefer.* **2017**, *62*, 257–261. [CrossRef]
21. Spence, C.; Wan, X. Beverage perception and consumption: The influence of the container on the perception of the contents. *Food Qual. Prefer.* **2015**, *39*, 131–140. [CrossRef]
22. Youssef, J.; Juravle, G.; Youssef, L.; Woods, A.; Spence, C. Aesthetic plating: A preference for oblique lines ascending to the right. *Flavour* **2015**, *4*, 27. [CrossRef]

23. Simmons, J.P.; Nelson, L.D.; Simonsohn, U. False-positive psychology: Undisclosed flexibility in data collection and analysis allows presenting anything as significant. *Psychol. Sci.* **2011**, *22*, 1359–1366. [CrossRef] [PubMed]
24. Simmons, J.P.; Nelson, L.D.; Simonsohn, U. Life after p-hacking. *SSRN Electron. J.* **2013**. [CrossRef]

 *beverages*

Article

Assessing the Impact of Closure Type on Wine Ratings and Mood

Charles Spence * and Qian (Janice) Wang

Crossmodal Research Laboratory, Department of Experimental Psychology, University of Oxford, South Parks Road, Oxford OX1 3UD, UK; qian.wang@psy.ox.ac.uk
* Correspondence: charles.spence@psy.ox.ac.uk

Received: 7 October 2017; Accepted: 27 October 2017; Published: 3 November 2017

Abstract: We report on a preliminary study designed to assess the impact of the sound of the closure on the taste of wine. Given that people hold certain beliefs around the taste/quality of wines presented in bottles having different closure types, we expected that the sound of opening might influence people's wine ratings. In particular, if participants hear a cork being pulled vs. the sound of a screw-cap bottle being opened then these two sounds will likely set different expectations that may then affect people's judgment of the taste/quality of the wine that they are rating. In order to test this hypothesis, 140 people based in the UK (and of varying degrees of wine expertise) rated two wine samples along four scales, three relating to the wine and one relating to celebratory mood. The results demonstrated that the sound of a bottle being opened did indeed impact ratings. In particular, the quality of the wine was rated as higher, its appropriateness for a celebratory occasion, and the celebratory mood of the participant was also higher following the sound of the cork pop. These results add to the literature demonstrating that the sounds of opening/preparation of food and beverage products can exert a significant influence over the sensory and hedonic aspects of people's subsequent tasting experience.

Keywords: closure type; opening sounds; wine perception; expectations; packaging

---

1. Introduction

A growing body of empirical research demonstrates that the sensory properties of the packaging in which drinks are presented can exert a significant influence over the ensuing tasting experience. This turns out to be true both when products are consumed direct from the packaging, but also when the product is first poured into a drinking vessel such as a glass, cup or mug [1]. Branding obviously plays a role here [2], but beyond that researchers have assessed the impact of everything from packaging/label color [3–5], through weight and other material properties [6,7]. One factor that has, until recently, received less empirical interest, though, relates to the sound of opening. It is our belief that auditory cues provide information that may help set certain expectations and hence color the subsequent tasting experience [8].

The general framework for interpreting such findings, showing as they do that the sensory aspects of the packaging influence the tasting experience, has been in terms of expectations and sensation transference. According to the former account, we normally generate expectations prior to tasting [9]. In terms of the expectations account, it is worth noting that expectations can be triggered by what we see, but also by what we hear, feel, smell, read etc. Those expectations, both sensory and hedonic, can then anchor the subsequent tasting experience. According to the sensation transference account, what we feel about the packaging, e.g., that it is high quality, weighty, traditional, etc. is then transferred to, or biases our ratings of, the contents [10].

In the present study, we were particularly interested in whether changing what people hear prior to tasting a wine would influence their sensory and hedonic expectations and hence their experience of a wine. In this regard, one of the most iconic opening sounds is the pop of the cork being removed from a wine bottle. According to Hallgarten [11], cork bottle stoppers made from the cork oak tree were first introduced towards the end of the 18th Century. Apocryphally, Dom Perignon, the cellarmaster at the Abbey of Hautvilliers, was one of the first to plug his bottles with cork bark. For those wine drinkers who believe that wine from cork-stoppered bottles tastes better [12], we hypothesized that they would give a wine a higher quality rating after hearing the pop of a cork, compared to after hearing the crack of the screw cap bottle [13][1]. In the present study, we therefore investigated whether the sound of a cork-stoppered bottle being opened (i.e., the popping sound) versus the sound of a screw-cap bottle being opened would influence people's ratings of a wine, and/or their celebratory mood. The experiment was conducted in two parts. Initially, the participants heard only the sound of bottle opening and rated two separate glasses of wine. In the second part of the study, the participants actually opened the cork/screw-cap bottle and poured themselves a glass of wine from each bottle type before rating them. This aspect of the experimental design allowed us to assess the multisensory contributions of actually opening and pouring the wine from either type of bottle (cork-closure vs. screw-cap) rather than just hearing the sound. Product experience is typically multisensory (involving sight, sound, touch, haptics, etc. [14]), and one might expect that the very action of pulling the cork would have more of an effect on ratings than merely hearing the sound. By contrast, however, if it is just the knowledge (or, better said, belief) that a certain wine came from a bottle with a particular type of closure that is doing the work, it may not matter whether that belief is based on unisensory (i.e., auditory) or multisensory (auditory plus haptic) cues.

The ratings were separated into questions concerning the quality of the wine (How intense? How would you rate the quality?) and the context/situation/mood around consumption (How appropriate is the wine for celebration? How much of a celebratory mood do you feel in right now?). The reason being that previous research suggests that the impact of product-extrinsic sounds such as sonic seasoning (i.e., musical compositions selected to accentuate a certain taste in a food) may be mediated, at least in part, by emotional valence [8,15]. Hence, in the present study, we wanted to try and ascertain by which route(s) people's ratings of the wines may have been affected.

2. Methods

Participants. A convenience sample of 140 participants was tested in the experiment (92 male and 47 female, 1 non-response). There were 115 right-handers, 19 left-handers, and 6 non-response. The participants indicated which age bracket they fell into (see Table 1). The participants also rated their knowledge of wine on a 1–5 scale (see Table 2) from 1—Novice to 5—Expert.

Table 1. The age distribution of participants.

Age Bracket	Number of Participants
18–25	22
26–35	61
36–45	29
45+	25

---

[1] It is worth noting here, that people's associations around the quality of screw-top bottles has changed in recent years, with some New World producers starting to change people's mind-set, no longer necessarily associating screw-cap with a lower quality product. Note that there has been something of a similar battle raging in the craft beer market between bottle and can format [1].

Table 2. The wine knowledge distribution of participants.

Wine Knowledge	Number of Participants
1 (Novice)	8
2	36
3	67
4	17
5 (Expert)	10

Design and Procedure. The experiment was split into two parts. In each part, the participants were given two wines to taste and rate sequentially. In the first part of the experiment, the participants rated one glass of wine after hearing the sound of a cork popping out of the bottle, and the other after hearing the sound of a screw-top bottle being opened. In the second part of the study, the participants actually opened a screw-top bottle themselves and rated the wine from that bottle and then pulled the cork on a stoppered wine bottle and rated the other wine. The bottle labels were obscured. In total, the participants gave ratings on four scales (see Table 3) for each of the four wines that they tasted. The two wines were *Terrazas de los Andes, Malbec 2015, Argentina* and a *Catena, Malbec 2015, Argentina*. Each wine was associated with the same closure type in each part of the study. Though, to be absolutely clear, the wine-closure pairing was counterbalanced across days/participants. The two wines were chosen to be similar but not identical and hence to make sure that there was a meaningful difference between the wines in each of the two parts of the experiment. The wine that was associated with each opening sound/opening action was counterbalanced across participants on the different days on which the study was conducted. For 91 of the participants (39 Press (including journalists and wine industry reporters) plus 52 participants on Day 1), the cork opening sound and the cork bottle opening were associated with the *Terrazas de los Andes* wine while the sound of the screw-cap opening and the action of opening and pouring from a screw-top bottle with the *Catena*. This wine-condition match was reversed for the remaining 49 participants (who participated on Day 2). The cork opening always preceded the screw-cap opening. Finally, the participants were asked "Do you prefer to buy a bottle of wine sealed with a cork or a screw cap?" The whole experiment took approximately 5–10 min to complete per person.

Table 3. The four ratings that participants were asked to give for each wine.

	Least				Most
How do you rate the intensity of the wine?	1	2	3	4	5
How do you rate the quality of the wine?	1	2	3	4	5
How appropriate do you think this wine is for a celebration?	1	2	3	4	5
How much of a celebratory mood do you feel in right now?	1	2	3	4	5

3. Results

In response to the question, "Do you prefer buying a bottle of wine sealed with a cork or a screw cap?" 113 responded 'cork', 13 responded 'screw-cap', and 14 failed to respond. A Chi squared test of goodness of fit revealed a significant preference for cork closures, $X^2(3140) = 142.89, p < 0.001$. This result is perhaps not surprising given that Argentine Malbec, and red wines in general, are typically sealed with a cork closure in the UK, where the study was conducted.

Pearson's correlations were calculated between the four measures of intensity, quality, celebration appropriateness, and celebratory mood. There were significant positive correlations between all pairs of measures (see Table 4). Consequently, a multivariate analysis of variance (MANOVA) was conducted with experimental condition (sound only or sound + touch), wine type (*Terrazas de los Andes* or *Catena*), and closure type (cork or screw-cap) as the between-participant factors. See Table 5 for average values of the four measures.

Table 4. Pearson correlation coefficients between wine ratings, over all conditions. * Indicates correlations that were significant at $p < 0.01$.

	Intensity	Quality	Celebration Appropriateness	Celebratory Mood
Intensity	1	0.34 *	0.19 *	0.15 *
Quality		1	0.62 *	0.31 *
Celebration appropriateness			1	0.49 *
Celebratory mood				1

Table 5. Mean ratings of Bonferroni-corrected pairwise comparisons between conditions (standard error in parentheses). Bold typeface indicates significant difference between cork and screw-cap conditions ($p < 0.05$).

INTENSITY		
SOUND ONLY	cork	screwcap
Terrazas de los Andes	3.50 (0.09)	3.49 (0.12)
Catena	3.14 (0.12)	3.07 (0.09)
SOUND + TOUCH	cork	screwcap
Terrazas de los Andes	3.41 (0.09)	3.27 (0.13)
Catena	2.98 (0.13)	2.89 (0.09)
QUALITY		
SOUND ONLY	cork	screwcap
Terrazas de los Andes	3.30 (0.09)	3.25 (0.13)
Catena	**3.35** (0.13)	**2.90** (0.09)
SOUND + TOUCH	cork	screwcap
Terrazas de los Andes	3.29 (0.09)	3.10 (0.13)
Catena	**3.40** (0.13)	**2.98** (0.09)
CELEBRATION-APPROPRIATE		
SOUND ONLY	cork	screwcap
Terrazas de los Andes	3.04 (0.10)	3.02 (0.14)
Catena	**3.31** (0.14)	**2.86** (0.11)
SOUND + TOUCH	cork	screwcap
Terrazas de los Andes	3.25 (0.10)	3.04 (0.14)
Catena	**3.35** (0.14)	**2.80** (0.11)
CELEBRATORY MOOD		
SOUND ONLY	cork	screwcap
Terrazas de los Andes	**3.32** (0.11)	**2.86** (0.15)
Catena	3.31 (0.15)	2.99 (0.11)
SOUND + TOUCH	cork	screwcap
Terrazas de los Andes	3.37 (0.11)	3.10 (0.15)
Catena	3.25 (0.15)	3.03 (0.11)

Overall, there was a significant main effect of closure type ($F(4542) = 4.83$, $p = 0.001$, *Wilk's* $\lambda = 0.97$; see Figure 1). Further univariate tests revealed that closure type significantly influenced the perceived quality of the wine ($F(1542) = 12.44$, $p < 0.0005$, $\eta_p^2 = 0.022$), celebration appropriateness ($F(1542) = 12.27$, $p < 0.0005$, $\eta_p^2 = 0.022$), and the celebratory mood of the participant ($F(1542) = 11.86$, $p = 0.001$, $\eta_p^2 = 0.021$). More specifically, wines with cork closures (or merely accompanied by the sound of a cork-stoppered bottle being opened) were rated as higher in quality ($M_{cork} = 3.33$, $SE = 0.06$, $M_{screwcap} = 3.06$, $SE = 0.06$), more appropriate for celebrations ($M_{cork} = 3.24$, $SE = 0.06$, $M_{screwcap} = 2.93$, $SE = 0.06$), and induced more of a celebratory mood ($M_{cork} = 3.31$, $SE = 0.06$, $M_{screwcap} = 3.00$, $SE = 0.06$). However, there were no significant effect of closure type on ratings of wine intensity ($F(1542) < 1$, *n.s.*).

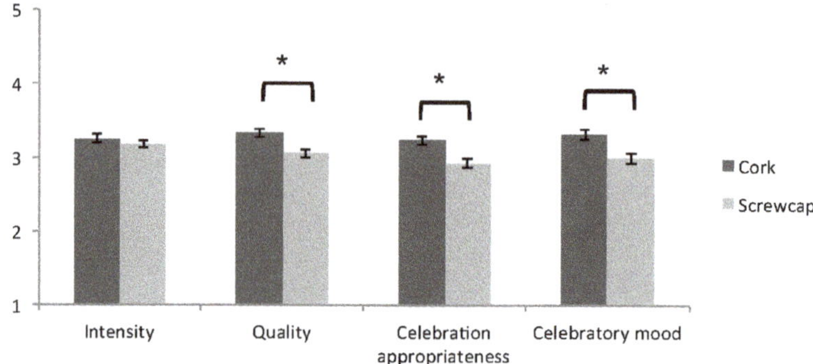

Figure 1. Mean values of wine intensity, wine quality, celebration appropriateness, and celebratory mood for wines with cork versus screw-cap enclosures. Error bars indicate standard error of means, and asterisks (*) indicate significant differences at $p = 0.05$. The y-axis reflects mean ratings on the 5-point scale.

In addition, there was a main effect of wine type ($F(4542) = 6.67$, $p < 0.0005$, *Wilk's* $\lambda = 0.95$). Further univariate tests revealed the effect of wine type on ratings of intensity ($F(1542) = 26.00$, $p < 0.0005$, $\eta_p^2 = 0.046$), with the *Terrazas de los Andes* wine being perceived as more intense than the *Catena* ($M_{Terrazas} = 3.43$, $SE = 0.05$, $M_{Catena} = 3.00$, $SE = 0.05$). This latter result may help to explain why larger differences were seen in ratings of quality for *Catena* than for the *Terrazas de los Andes* wine (i.e., there was more room for improvement in the former case).

Finally, it is worth noting that there was no effect of experimental condition ($F(4542) = 1.48$, $p = 0.21$). In other words, it did not seem to matter whether the knowledge of wine enclosure was based on unisensory (hearing alone) or multisensory (hearing plus touch) cues.

4. Discussion and Conclusions

The vast majority of those questioned in the present study reported that they preferred the taste/flavor of the wine from a cork-stoppered bottle. One explanation for this preference emerges from the analysis of the results of the wine ratings: On average, the 140 participants tested in the present study rated the wine that they themselves served from a cork-stoppered bottle, or that they heard being poured from a cork-stoppered bottle, as being of higher quality, as more appropriate for celebrations, and as inducing a greater celebratory mood [16]. By contrast, there was no significant effect of closure type on ratings of wine intensity. These results are consistent with the view that the effect on mood—rather than any changes in the perception of the wine itself—might be driving part of the change in the ratings elicited by the sound of the cork, as opposed to the screw-cap closure [15,17]. The results of the present study further suggest that it is knowledge about the closure type, rather than how that knowledge was acquired (i.e., just from sound versus from sound plus touch—sight and action of opening and pouring the wine) that led to the ratings differences reported here. The suggestion is that distinctive packaging sounds, just as for the other aspects of the packaging, can set expectations in the mind of the consumer.

In future research, it would be interesting to try and replicate these findings using a between-participants experimental design (rather than just a mixed-participants design, as conducted here). This would help address one potential concern associated with the latter design, namely, that it draws attention to the closure type in a way that might not necessarily be observed in everyday life. If possible, it may also be worth conducting future research under conditions that are more naturalistic (ecologically valid) and more conducive to celebration than the experimental setup used here. It may also be advantageous to randomize the order in which ratings are made in order to prevent any order/sequential effects from coloring the results. Beyond that, it would obviously be beneficial to randomize the order in which the cork and screw-cap bottle sounds/actions were presented.

Of course, it should be noted that the effect size of the closure type differences on various ratings—as reported by partial eta squared values—are small [18,19]. In a practical sense, therefore, the decision to go with cork or screw-cap on the part of the manufacturer may not be expected to result in a drastic change in the tasting experience. Nevertheless, the fact that an overwhelming proportion of participants (113 vs. 13) reported preferring to buy a cork-sealed bottle of wine over a screw-cap bottle implies that small differences in perceived attributes of a wine may nevertheless still shape buying decisions [20].

Finally, it is important to note that there is also a temporal aspect to people's feelings about different closure types: That is, people's views likely changes over time. Indeed, as the years pass, there is a sense that screw-top bottles are starting to be associated with a better quality of product than they once were. There may also be relevant cross-cultural variations in this regard too; for instance, people from regions where screw-cap enclosures are more prevalent (e.g., Australia or New Zealand) might not necessarily show a preference for corks over screw-caps. Hence, the present results (showing a clear preference for wines served from a cork-stopped bottle) should not, at least for the moment, be generalized beyond the mostly UK-based participants[2] who took part in the present study in 2017 and who rated two Argentinian red wines.

Nevertheless, having raised these various caveats/concerns, the key point remains that these results provide the first empirical demonstration that one and the same wine is rated more highly in terms of its quality, it may be rated as more appropriate for celebrations, and also induces a greater celebratory mood when served in a cork-stopped rather than a screw-cap closure bottle. While such claims are certainly not new, the original part of this study resides in showing the effects can be elicited by nothing more than the sound of the closure. These results help to emphasize the importance of the packaging, and 'the image mold', to our experience of the contents [14].

Ultimately, though, these psychological benefits to the tasting experience associated with the cork closure should be weighed against the cork taint that affects some small proportion of cork-stopped bottles. Cork taint is associated with the presence of 2,4,6-trichloroanisole (TCA), which occurs as a reaction between a penicillium mold present in the cork and the chlorinated sterilants [21–23]. Here, there are several important points to note. First, sensitivity to this fault varies widely across the population [24]; it also varies as a function of the style of wine [25]. Second, innovations in enology mean that the incidence of this problem is much lower today than it has been in the past [12]. Third, it is important to note that cork taint also affects other closure types too [12] and can be transmitted to a wine via contaminated winery equipment other than cork [21]. Nevertheless, despite any potential technical limitations, it is clear that the sound of a cork-stopped wine bottle being opened conveys a psychological impact that can help to enhance the ratings and experience of anyone who hears it.

Acknowledgments: Funding for this study was provided by Bompas & Parr. No funds were received to cover the costs of publishing in open access.

Author Contributions: C.S. conceived and helped design the study; Q.W. analyzed the data from the study; C.S. and Q.W. wrote the paper.

Conflicts of Interest: C.S. has worked with APCOR and Clarion Communications on the psychological associations with cork closures. Q.W. declares no conflict of interest.

References

1. Barnett, A.; Velasco, C.; Spence, C. Bottled vs. canned beer: Do they really taste different? *Beverages* **2016**, *2*, 25. [CrossRef]

---

2   Although US-based participants showed a similar preference for cork [14].

2. Gates, P.W.; Copeland, J.; Stevenson, R.J.; Dillon, P. The influence of product packaging on young people's palatability ratings for RTDs and other alcoholic beverages. *Alcohol Alcohol.* **2007**, *42*, 138–142. [CrossRef] [PubMed]

3. Barnett, A.; Spence, C. Assessing the effect of changing a bottled beer label on taste ratings. *Nutr. Food Technol.* **2016**, *2*. [CrossRef]

4. Cutler, L. Wine label design: What makes a successful label. *Wine Business Monthly*, 15 August 2006.

5. Lick, E.; König, B.; Kpossa, M.R.; Buller, V. Sensory expectations generated by colours of red wine labels. *J. Retail. Consum. Serv.* **2017**, *37* (Suppl. C), 146–158. [CrossRef]

6. Kampfer, K.; Leischnig, A.; Ivens, B.S.; Spence, C. Touch-taste-transference: Assessing the effect of the weight of product packaging on flavor perception and taste evaluation. *PLoS ONE* **2017**, *12*, e0186121. [CrossRef] [PubMed]

7. Piqueras-Fiszman, B.; Spence, C. The weight of the bottle as a possible extrinsic cue with which to estimate the price (and quality) of the wine? Observed correlations. *Food Qual. Preference* **2012**, *25*, 41–45. [CrossRef]

8. Spence, C.; Zampini, M. Auditory contributions to multisensory product perception. *Acta Acust. United Acust.* **2006**, *92*, 1009–1025.

9. Piqueras-Fiszman, B.; Spence, C. Sensory expectations based on product-extrinsic food cues: An interdisciplinary review of the empirical evidence and theoretical accounts. *Food Qual. Preference* **2015**, *40*, 165–179. [CrossRef]

10. Spence, C.; Gallace, A. Multisensory design: Reaching out to touch the consumer. *Psychol. Mark.* **2011**, *28*, 267–308. [CrossRef]

11. Hallgarten, F. *Wine Scandal*; Sphere Books: London, UK, 1987.

12. Goode, J. *Wine Science*; Mitchell Beazley: London, UK, 2005.

13. Spence, C.; Wang, Q.(J.). Sensory expectations elicited by the sounds of opening the packaging and pouring a beverage. *Flavour* **2015**, *4*, 35. [CrossRef]

14. Spence, C. Multisensory packaging design: Color, shape, texture, sound, and smell. In *Integrating the Packaging and Product Experience: A Road-Map to Consumer Satisfaction*; Chen, M., Burgess, P., Eds.; Elsevier: Oxford, UK, 2016; pp. 1–22.

15. Wang, Q.(J.); Spence, C. Assessing the role of emotional associations in mediating crossmodal correspondences between classical music and wine. *Beverages* **2017**, *3*, 1. [CrossRef]

16. Marin, A.B.; Jorgensen, E.M.; Kennedy, J.A.; Ferrier, J. Effects of bottle closure type on consumer perceptions of wine quality. *Am. J. Enol. Vitic.* **2007**, *58*, 182–191.

17. Spence, C.; Wang, Q.(J.). Wine & music (II): Can you taste the music? Modulating the experience of wine through music and sound. *Flavour* **2015**, *4*, 33.

18. Cohen, J. *Statistical Power Analysis for the Behavioral Sciences*; Routledge Academic: New York, NY, USA, 1988.

19. Lakens, D. Calculating and reporting effect sizes to facilitate cumulative science: A practical primer for *t*-tests and ANOVAs. *Front. Psychol.* **2013**, *4*, 863. [CrossRef] [PubMed]

20. Mueller, S.; Szolnoki, G. The relative influence of packaging, labeling, branding and sensory attributes on liking and purchase intent: Consumers differ in their responsiveness. *Food Qual. Preference* **2010**, *21*, 774–783. [CrossRef]

21. Bird, D. *Understanding Wine Technology*; DBQA Publishing: Newark, UK, 2010.

22. Buser, H.-R.; Zanier, C.; Tanner, H. Identification of 2,4,6-trichloroanisole as a potent compound causing cork taint in wine. *J. Agric. Food Chem.* **1982**, *30*, 359–362. [CrossRef]

23. Duncan, B.C.; Gibson, R.L.; Obradovic, D. 2,4,6-trichloroanisole and cork production. *Wine Ind. J.* **1997**, *12*, 180–184.

24. Prescott, J.; Norris, L.; Kunst, M.; Kim, S. Estimating a "consumer rejection threshold" for cork taint in white wine. *Food Qual. Preference* **2005**, *16*, 345–349. [CrossRef]

25. Mazzoleni, V.; Maggi, L. Effect of wine style on the perception of 2,4,6-trichloroanisole, a compound related to cork taint in wine. *Food Res. Int.* **2007**, *40*, 694–699. [CrossRef]

Review

Two Decades of "Horse Sweat" Taint and *Brettanomyces* Yeasts in Wine: Where do We Stand Now?

Manuel Malfeito-Ferreira

Linking Landscape, Environment, Agriculture and Food Research Centre (LEAF),
Instituto Superior de Agronomia, University of Lisbon, Tapada da Ajuda, 1349-017 Lisboa, Portugal;
mmalfeito@isa.ulisboa.pt

Received: 19 February 2018; Accepted: 4 April 2018; Published: 10 April 2018

Abstract: The unwanted modification of wine sensory attributes by yeasts of the species *Brettanomyces bruxellensis* due to the production of volatile phenols is presently the main microbiological threat to red wine quality. The effects of ethylphenols and other metabolites on wine flavor is now recognized worldwide and the object of lively debate. The focus of this review is to provide an update of the present knowledge and practice on the prevention of this problem in the wine industry. *Brettanomyces bruxellensis*, or its teleomorph, *Dekkera bruxellensis*, are rarely found in the natural environment and, although frequently isolated from fermenting substrates, their numbers are relatively low when compared with other fermenting species. Despite this rarity, they have long been studied for their unusual metabolical features (e.g., the Custers effect). Rising interest over the last decades is mostly due to volatile phenol production affecting high quality red wines worldwide. The challenges have been dealt with together by researchers and winemakers in an effective way and this has enabled a state where, presently, knowledge and prevention of the problem at the winery level is readily accessible. Today, the main issues have shifted from technological to sensory science concerning the effects of metabolites other than ethylphenols and the over estimation of the detrimental impact by ethylphenols on flavor. Hopefully, these questions will continue to be tackled together by science and industry for the benefit of wine enjoyment.

Keywords: wine; spoilage; *Brettanomyces*; *Dekkera*; volatile phenols; off-flavors

---

1. Introduction

The yeast genus *Brettanomyces* has been related with the production and characteristics of English beers since the beginning of the XX century [1]. It has been under the attention of early yeast physiologists due to its unusual fermentation stimulation by oxygen, coined as the Custer's effect [2]. In wines, the first isolates were obtained by Krumholz and Tauschanoff in 1933 without any particular technological concern [1], and the investigation on the genus was relatively scarce until the mid-nineties of last century. To understand the impact of these findings on Brett research, Figure 1 shows the number of peer-reviewed references retrieved from the Scopus database (www.scopus.com, assessed on the 29 January 2018) using "brettanomyces or dekkera and wine" as search words in article titles, abstracts, and keywords from 1959 until 2017. The increase in the numbers coincided with the publication of seminal research by scholars at Bordeaux University [3]. In addition, the total number was 339, which is more than half of the references that use only "brettanomyces or dekkera" (a total of 498). Although involved in other fermented foods, beverages, and ethanol production, most publications recognize that the role played by the species *Brettanomyces bruxellensis* (anamorph of *Dekkera bruxellensis*) in red wine spoilage is due to the production of "horse sweat" taint in bulk or bottled wines. Moreover, its effects are particularly notorious in high quality red wines aged in costly oak barrels, which considerably increase the economic losses provoked by spoilage yeasts in the wine industry. Presently, this species is regarded as the main threat posed by yeasts to wine quality [4], surpassing the research interest on the *Zygosaccharomyces* genus, which is another dangerous wine spoilage yeast. To understand the significance of the relative value of these figures, when the Scopus search executed using the key-words "zygosaccharomyces and wine" and "zygosaccharomyces", the numbers were 128 and 1122, respectively.

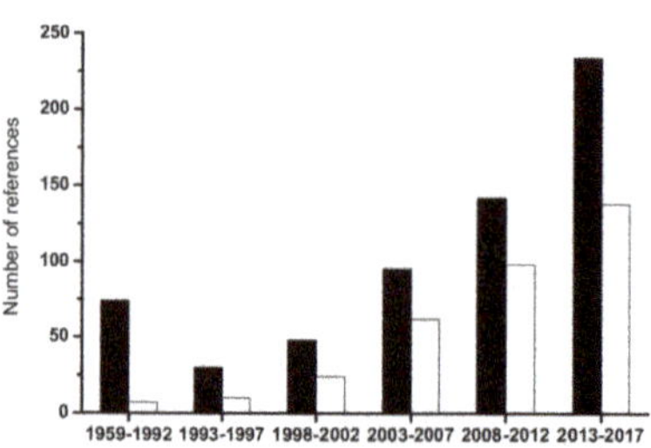

Figure 1. Number of references related with "*Brettanomyces*" (black bars) and with "*Brettanomyces* or *Dekkera* and wine" (white bars) retrieved from Scopus search engine (www.scopus.com).

The research of *Brettanomyces* gathers practically all fields of microbial research (physiology, metabolism, genomics, evolution, taxonomy, etc.) and has been regularly reviewed [1,5–13]. The aim of this review was to provide an additional approach that mainly concerns the most relevant scientific achievements and technological impacts, providing an update of the present knowledge and best practices on the prevention of this problem in the wine industry. Future challenges motivated by the sensory perception of volatile phenols will also be addressed because this issue is the subject of extensive debate among wine professionals and the general public.

2. Volatile Phenols (VPs): Their Incidence and Origin

The incidence of VPs in red wines is one of the most frequent wine defects alongside corkiness, reduced, and oxidized off-flavors [14]. The phenolic off-odors have been described as "medicinal", "phenolic", "rancid", "sweaty", "smoke", "Band-aid®", "barnyard" or "horse sweat". The most important VP is 4-ethylphenol (4-EP), followed by 4-ethylguaiacol (4-EG), and 4-ethylcathecol (4-EC). Earlier works by the team of Pascal Chatonnet during the 1990s provided values for sensory thresholds that are still a reference today. For instance, in Bordeaux red wines, the preference threshold for 4-EP is 620 µg/L, and for the mixture (10:1) of 4-EP and 4-ethylguaiacol is 426 µg/L [3]. However, other thresholds have been published that reflect the influence of other factors among which the wine matrix is essential [15]. For instance, wine body and oak flavor are like buffers of the tainting effect of volatile phenols. In addition, other products of Brett metabolism like isovaleric and isobutyric acids [11] also influence the perception of "horse sweat", which may explain why the smell of wines spiked with VPs is more objectionable than the smell of wines with the same concentrations but produced naturally.

The precursors of 4-EP, 4-EG, and 4-EC are hydroxycinnamic acids (*p*-coumaric, ferulic and caffeic acids, respectively), which are enzymatically decarboxylated by a cinnamate decarboxylase, leading to vinyl derivatives and reduced by a vinylphenol reductase, originating in the ethyl derivatives [11].

In grape juices those hydroxycinnamic acids are esterified, mainly to tartaric acid, in concentrations higher than 10 mg/L. In wines they may be present in the free or esterified form, either with tartaric acid [16,17], ethanol [18], hexoses [16] or polyphenols [7]. The release of hydroxycinnamic acids from anthocyanin esters during wine maturation may be only due to chemical reactions [7] but conversions of hydroxycinnamic acid precursors to VPs are typically dependent on enzyme or microbial activity. In grapes or grape juices, the tartaric esters may be hydrolyzed by enzymes from contaminant fungi or from commercial pectolytic preparations, both with cinnamoyl decarboxylase activity, which releases free hydroxycinnamic acid forms [11]. Ethyl and glucose esters of hydroxycinnamic acids may be metabolized by *Brettanomyces* [16,18] contrarily to tartaric acid esters [16,17]. Most of these tartaric esters are hydrolyzed only after malolactic fermentation, and it has been hypothesized that the hydrolytic activity of the lactic acid bacteria follows the completion of malic conversion to lactic acid [19,20]. Thus, the pool of hydroxycinnamic acids' precursors provides the substrate for the production of VPs that are higher than the preference thresholds depending on the presence of active *B. bruxellensis* populations [20,21]. *S. cerevisiae* may produce vinyl derivatives that may be later reduced by *B. bruxellensis*. This species is highly efficient in the conversion of *p*-coumaric acid with molar rates higher than 90% [22]. Then, the natural pool of hydroxycinnamic acids (>10 mg/L) in wines are not likely to be a limitation for the production of VPs higher than the preference thresholds. Therefore, all red wines may be affected by the taint, given that the yeasts are able to grow. The belief that there are some grape varieties more susceptible to Brett growth has no scientific evidence [23].

The absence of "horse sweat" in white wines is probably due to the easier Brett inactivation, given that hydroxycinnamic acids are also present in concentrations similar to those of red wines [24]. The yeast species *Pichia guilliermondii* and lactic acid bacteria (*Lactobacillus* spp., *Pediococcus* spp.) also have the ability to produce 4-EP, but their spoiling potential in wines is not comparable to that of *B. bruxellensis* [25–27].

In conclusion, the natural concentrations of hydroxycinnamic acids in wines are high enough to provide substrate for the production of volatile phenols much above the preference thresholds. Therefore, the key for the prevention of the problem is to understand the ecology and the behavior of *Brettanomyces* in wines in order to apply the adequate control measures and avoid, or reduce, the conversion of hydroxycinnamic acids to VPs.

3. Brett Ecology: Infection Routes in the Winery

The *Brettanomyces* yeasts are mostly associated with fermented products, particularly with the post-fermentation or aging period of alcoholic beverages, like wine, beer, cider, kombucha, and tequila. Some reports refer their isolation from cheeses or fermented milks. Besides the food industries, *Dekkera/Brettanomyces* spp. have also been reported in industrial ethanol fermentations. These yeasts are scarcely mentioned outside of fermentation environments. Occasional reports include isolation from olives, flours, carbonated beverages, and bees and air at ground level in fruit orchards, honeys, and tree exudates [5]. Being a fermenting species, *Brett* probably shares the same natural habitat as *S. cerevisiae* (e.g., soil, oak bark, decaying vegetal tissues, and tree exudates) but it is even rarer. Despite the preference for fermented products, *Dekkera/Brettanomyces* spp. are usually not dominant and are reported in a low percentage of analyzed samples. Their occasional detection in sparkling wines may be related with their resistance to carbon dioxide, which is higher than that of *S. cerevisiae* and *Z. bailli* [28]. Thus, Brett can be regarded as sporadic contamination yeast that appears in high numbers mainly when other microorganisms have been inhibited.

The dissemination of *B. bruxellensis* in wine related environments is hard to evidence from sources contaminated by other yeasts due to its low growth rate. As a consequence, the use of selective media and long incubation periods are essential to its recovery [29], although even when using selective media, it has been rarely isolated from grapes [30] and clean winery environments. Rotten grapes appear to bear higher populations, which most likely explains the higher incidence of phenolic wines in vintages affected by this problem [31]. The frequency of isolation from cellar equipment increases in the presence of juice, wine residues, and leftovers (husks and pomaces). Insects and air dusts are probable vehicles of dissemination among infecting spots [32]. In contrast, it is a common contaminant easily recovered from red wines where numbers may attain high levels (>10^4 cells/mL) if preventive measures are not taken [33].

The frequency of isolation in red wines explains why careful vigilance must be given to wines purchased from other sources or sold as bulk wines to other wineries. Once established in a winery, Brett is very difficult to eradicate, especially when it has contaminated the wooden vats.

Old Wooden Vats: A Well-Known Ecological Niche

The traditional wooden vats used for red wine, sherry, lambic beer or cider production are typical niches of *B. bruxellensis*. In wines, the onset of problems related with its activity coincided with the worldwide increase in the utilization of oak barrels during the last 25 years. However, wines matured in stainless steel tanks are also affected by these yeasts. The survival in barriques is facilitated by the diffusion of oxygen either by stimulating yeast growth [34] or by reducing the levels of molecular sulfite active against yeasts. In addition, it is believed that cell immobilization in the wood structure contributes to the protection against preservatives. The ability to form pseudomycelium might as well favor the colonization of the porous structure of the wood and of the spaces between the staves and the grooves. The wood of new barriques is not the source of Brett but it is readily contaminated if filled with infected wine. On the contrary, high ethanol wines (>17% *v/v*) are not susceptible to these yeasts even if aged in old wooden vats, such as in the case of sherry or port-style wines.

Overall, the incidence of Brett increases during grape processing, from fermentation to red wine ageing. The routes of contamination may not be perfectly known but this yeast easily finds the way to infect stored red wines and initiates spoilage if not stopped in due time.

4. Brett Behavior and Tolerance in Wines

The practical absence of *Brettanomyces* from the early stages of grape processing is no longer observed after the end of malolactic fermentation. In certain cases its activity may already begin in the period between alcoholic and malolactic fermentations. It is not as tolerant to ethanol or sulfite as *S. cerevisiae* or *Zygosaccharomyces bailii* [35] but it has the ability to remain viable for long periods. The increase in their predominance appears to be the result of an exceptional resistance to minimal nutrient conditions, which is seen as the determinant in their survival during the production and storage steps and in their development once the environment becomes favorable (e.g., reduction in free sulfite during ageing) [11,30]. For instance, viable *Brettanomyces* have been reported in red wine bottled for more than 50 years [33].

4.1. Attention to the Unnoticed Presence of Brett

The behavior of *D. bruxellensis* in co-culture with *S. cerevisiae* (molasses, grape juice, and synthetic medium) is characterized by showing null or slow growth until about the end of fermentation. In post-fermentation it may grow, attaining levels as high as those observed with *S. cerevisiae* [22]. After inoculation in wines, *D. bruxellensis* growth shows a typical bell-shaped curve, with exponential growth followed by cell death. When the stress imposed to cells is high, a sharp decrease in viability is frequently observed and the wine seems to be Brett-free [24]. In fact, cells may be in a viable but non-culturable state (VBNC cells) [36]. However, these observations may well result from the fact the sample volume analyzed was too low to recover viable cells. Whatever the explanation, the technological significance of these observations is that after an apparent Brett-free period the growth may be re-initiated either by surviving or VBNC cells, as shown in Figure 2.

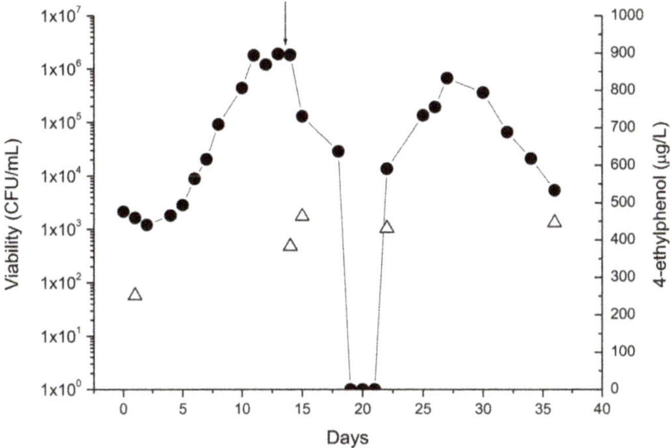

Figure 2. Effect of chitosan on the viability of growing cells *D. bruxellensis* (•) and on 4-ethylphenol (△) production. Arrow: moment of chitosan addition. Absence of viable cells (CFU <1/mL) is indicated as 1 CFU/mL due to the logarithmic scale of the *y*-axis.

4.2. Resistance to Antimicrobials

Sulphur dioxide is the most common and effective preservative utilized in wineries. However, levels may be as high as 40 mg/L of free sulfite at pH 3.5, to control *D. bruxellensis* in wines aged in barrels [24]. The active form is the molecular one, and so the lower the pH the higher the proportion of molecular sulfite. The effective free sulfite values in red wines may look somewhat high but probably reflect the proportion of sulfite bound to the anthocyanins that are counted as free sulfite by the current titration methods. The strategy should be directed to increase the ratio of free to bound sulfite so that inhibitory levels may be reached under a lower concentration of total sulfur dioxide.

Attempts to reduce sulfite utilization boosted the search for commercial alternatives. Sorbic acid has long been known to act against fermenting yeasts in bottled wines but *D. bruxellensis* is resistant to the maximum legal concentration of 200 mg/L in wines. Presently, the focus has been turned to dimethyldicarbonate (DMDC) and chitosan, which are now commercially available. Other antimicrobials, like killer proteins and peptides of microbial origin (e.g., zymocins) are reported as effective in research articles but are not yet spread in the industry [37,38].

DMDC is an effective agent against Brett growth, depending on the initial cell concentration [39] (Table 1). In wines matured in barrels it is an efficient tool to prevent blooms, being used in regular additions up to the maximum permitted level of 200 mg/L. In the EU it is only authorized just before bottling, in wines with more than 5 g/L of sugar, and it is claimed to be efficient if yeast counts that are less than 500 CFU/mL. The efficiency also depends on adequate DMDC homogenization, which requires the use of specific equipment that makes the treatment cost high (>0.05 €/bottle). In addition, DMDC hydrolysis releases methanol that should be monitored if excessive concentrations are suspected.

Table 1. Minimum lethal concentration of dimethyldicarbonate (DMDC) (mg/L) against several wine related microbial species as a function of initial cellular inoculum (adapted from [39]).

Species	500 cells/mL	>10^4 cells/mL	Species	500 cells/mL	>10^4 cells/mL
D. bruxellensis ISA 1791	100	300	*S. pombe* ISA 1190	100	>300
P. guilliermondii ISA 2105	100	300	*Z. bailii* ISA 1307	25	200
S. cerevisiae ISA 1000	100	200	Lactic acid bacteria	>300	>300
S. cerevisiae ISA 1026	100	200	Acetic acid bacteria	>300	>300

Chitosan is a natural derivative of chitin and prevents Brett growth in wines with variable efficiency [40]. As with sulfite or DMDC, the important is to ensure that cells do not recover viability after the death phase as shown in Figure 2. The legal limit (0.1 g/L to inactivate Brett) is enough when contaminations are low but it is not as effective when cells are present in high numbers and growing. The product is not cheap (>0.05 €/L) and careful examination of the balance between costs and benefits should be done before application.

4.3. The Hurdle Concept in Food and Wine Preservation

The judicious utilization of sulfite is the key to preventing the building up of *D. bruxellensis* contaminations, however, sometimes the edge of its OIV guideline legal limit of utilization in red wines is reached (150 mg/L in red wines with less than 4 g/L reducing sugar, 300 mg/L in wines with more than 4 g/L reducing sugar) (http://www.oiv.int, accessed on the 11 February 2018). Then, DMDC or chitosan may be used to minimize the utilization of sulfur dioxide. These chemical preservatives are not the only options. The winemaker may also choose to control Brett using physical methods. The overall concept of spoilage prevention is known as the *hurdle concept* in food microbiology. The idea is to weaken microbial populations by making them to "jump" several hurdles. The more hurdles to jump the easier it would be to prevent microbial growth. In wineries, these hurdles include environmental factors (e.g., storage temperature, dissolved oxygen) and processing factors (e.g., fining, filtration, heat treatments, high pressure, pulse electric fields, preservatives) [41–43] that when applied correctly contribute to decreasing the utilization of sulfur dioxide.

In conclusion, Brett inactivation may be achieved by several alternative processes. The effectiveness of the different options must be ascertained under each real condition by appropriate monitoring strategies.

5. Brett Prevention: How to Monitor Contaminations

Wine technologists have two different attitudes when facing the threat of *D. bruxellensis*. One, which we can call "optimistic", results from the absence of wines spoiled by the yeasts and the thought that it only happens to the others. The other, which we can call "pessimistic", results from traumatic experiences in the past. Naturally, neither are the correct attitudes. The first runs under high risks that will, eventually, bring disastrous consequences. The second leads to exaggerated precautions and to high costs, either economically or in wine quality. Assuming that these yeasts are always present, although not detected, it is necessary to learn how to live with them in the winery and apply the best preventive measures.

5.1. To Know Where They Are and How Fit They Are

The first step to live with Brett is to know where it is. The proper prevention of Brett activity depends on its detection by microbiological analysis and to decide the most adequate treatment. Today, several microbiological methods are available to monitor *B. bruxellensis* periodically, including highly specific molecular methods [44–46]. The most common and suitable technique is plate counting. However, many wineries do not have regular microbiological control. The small dimension of wine enterprises is a serious limitation to the development of a routine microbial control. The costs of equipment from the simplest for plate counting to advanced instruments (e.g., fluorescence microscope, real-time polymerase chain reaction (PCR), flow cytometer) and the requirement for skilled labor are still a burden for most small and medium enterprises. Therefore, the economic losses associated with *D. bruxellensis* activity have moved many companies to ask for external support, which may be easily found today.

In our lab routine, we currently apply a simplified detection technique based on growth in a selective solid or liquid medium. Serial dilutions of wine samples are inoculated in plates or test tubes, incubated at room temperature and the results checked after 4 to 15 days. These microbial determinations are accompanied by 4-EP determination and gas chromatography as a measure of the spoiling activity.

The microbiological analysis should be regarded at two levels: (i) when the wines are stored in bulk, in oak barrels or in tanks; and (ii) when the wine is to be bottled. In the first situation, the main purpose is to avoid the production of VPs in levels high enough to produce off-flavors and off-tastes. Thus, it is not mandatory to eliminate Brett completely, but to assure that the level of contamination or of activity is low enough to keep 4-EP levels constant. In the second situation, the main purpose is to have bottled wine free from these yeasts. Only one viable cell per bottle may be the cause for spoilage much later.

5.2. Microbial Guidelines

In bulk stored wines, it is satisfactory to detect *B. bruxellensis* monthly, bimonthly or even every three months, depending on the contamination history. The sample volumes are from 100, 10, 1, to 0.1 mL, from a blend composed by wine from the interface air/liquid and from different depths of the container. In case the result is positive for 1 mL, or less, and the level of 4-EP is higher than 150 µg/L, it is recommended to reduce the microbial populations by fine filtration (<1 µm) immediately, accompanied by sulfite addition [35]. Negative results in 1 mL mean that contamination is low and only fining followed by sulfite should be enough. Chitosan, when contamination is low, or thermal treatments, when numbers are high, are two of the alternatives during wine ageing. The effectiveness of the treatments must be ascertained by microbiological analysis. When 4-EP levels are stable there is no need to reduce contamination because cells may be present but are not active, thus reducing the additions of sulfite during storage.

For wines before bottling, the criteria are more stringent, and detection should be made on 100, 10, and 1 mL of wine, sampled as described above. In case the result is positive in 1 or 10 mL, it is recommended a very fine or sterilizing filtration. If positive detection is only obtained for 100 mL, it is admissible to control viable cells only by the addition of preservatives (e.g., 40 mg/L of free sulfite, at pH 3.50). In this case, bottling must be technically correct and dissolved oxygen should be lowered to practically zero. Otherwise, it may be recommended to use DMDC or, as an alternative, a thermal treatment to destroy viable cells [35].

5.3. The Question of Real Time PCR

Given the slow growth of these yeasts, cultured media can only give results after more than 4–5 days and so early detection depends on the use of direct techniques. Presently, there are several real time PCR protocols that provide results in about 4–6 h and have a high sensitivity (<10 cells/mL) [45,46]. They have two main drawbacks. One is the cost, which is very high for a routine analysis (>60 €/sample). The second is related with false positive responses given by the DNA of dead cells. In this case, protocols must be adapted to remove this DNA from the samples [47].

6. Brett Prevention: How to Control, Kill, and Cure

Knowing the monitoring and treatment options available, we will describe below our main observations and decisions during our empirical experience in wineries. The first idea is always to reduce cell numbers and growth so that treatments may be limited as much as possible. If these measures are not taken in due time, VPs may attain high levels and curative measures may be used.

6.1. Reduction of Dissemination

The infections come into the winery through the grapes, wines, insects, and used barriques. In the case of grapes, it is not efficient to analyze and separate the infected grapes. Assuming that the prevalence of *D. bruxellensis* is higher in vintages of poor sanitary quality grapes, care should be taken to (i) minimize the time between alcoholic and malolactic fermentation and (ii) avoid cross contaminations that may jeopardize other finished wines. Outsourced red wines must be monitored before blending and treated adequately (see below). Insect dissemination through winery atmosphere, during harvest or during bottling, must be minimized. New barriques do not contain *Brettanomyces*, but used barriques, when recovered, are a most probable infection source for the barreled wines. However, new barriques may be even more suitable to support Brett growth because of higher oxygen permeability and nutrient release from the wood [6].

Once inside the winery, the most important factor to reduce dissemination is proper hygiene to avoid cross contaminations when moving wine. Common disinfectants used in the food industry are effective against yeast species and also against *Brettanomyces* [48]. The most efficient products are alkaline detergents, iodophors, and peracetic acid based sanitizers. Thus, the main concern is how to sanitize, properly, points of complex geometry or difficult access, particularly in bottling machines, like dead ends of filters, valves, gauges, or hoses. Pumps and hoses for wine transfer between tanks is another concern. Less efficient, or virtually impossible, is the sterilization of wooden vats common in traditional fermentation processes and in modern fashionable wine ageing. Sanitation with hot water or steam is essential, although not completely efficient. Even after steaming, *Brett* may be recovered from wood layers up to 4–6 mm below the surface [49]. Whatever the treatment adopted in wineries, one should bear in mind that, in barrels, the critical factor is the inability of the disinfecting agent (e.g., hot water, steam, ozone, microwaves, UV light) to reach the deeper layers of the wood.

6.2. Prevention of D. bruxellensis Growth and 4-EP Production

The prevention first depends on careful *D. bruxellensis* monitoring during all wine storage time. It should begin after malolactic fermentation or even before, when extended periods occur between the end of wine fermentation and the onset of malolactic fermentation. During this period wines are left unprotected by sulfite and kept at higher temperatures to promote the bioconversion of malic acid, which stimulate premature Brett growth. After malolactic fermentation, the frequency of analysis depends on its detection, but even in the absence of contaminations at least three analyses per year (after winter, before harvest, after harvest) should be done to avoid unnoticed growth [35]. This frequently happens during the harvest period, when temperatures are higher and attention is directed to pick grapes and ferment wines.

The main concern when a contaminated sample appears is to avoid any cross contamination. This is especially important when processing products from external sources, such as purchased wines of unknown origin. The best solution, at first, is to detect the presence of *D. bruxellensis* to determine the measures to be taken. In case wines are highly contaminated, the most efficient measure would be to filter through pores tighter than 1.0 μm, or even 0.8 μm, given that sulfur dioxide reduces cell size [50]. Even knowing the difficulty to sterilize by filtration young wines and the controversy of such measures, our opinion is that it should be considered, particularly when wines are to be matured in costly oak barrels.

Concerning factors promoting wine colonization, special attention should be paid to the levels of free sulfite, levels of dissolved oxygen, presence of residual sugars, and the storage temperature. Ideally, if wines could be stored at less than 10 °C there would be no yeast growth and concurrently no noticeable 4-EP production (Figure 3).

Oxygen is essential to wine ageing but it also stimulates growth and the production of volatile phenols [34]. Even under the practical absence of oxygen *D. bruxellensis* grows and produces 4-ethylphenol but at lower rates that are enough to affect wine quality. Therefore, all operations contributing to oxygen diffusion must be minimized, or carefully monitored, like rackings, pumpings, toppings, bottling and, particularly, micro-oxygenation—a fashionable process to accelerate red wine ageing. Oxygen also contributes to a quicker loss of free sulfites during storage, opening a window of opportunity for Brett growth. In our experience, the most frequent reason for unexpected growth is the drop of free sulfites below 20 mg/L (at pH 3.5).

Another aspect to keep in mind is the level of nutrients in the wine. These yeasts may grow well under very low levels of residual sugar (<2 g/L and in wines where no nutrients were added).

However, reducing nitrogen additions to the minimum required to finish fermentations is always a wise strategy to minimize Brett growth during storage. The idea is to create a "nutrient desert" that turns the wine less susceptible to yeast growth. Most likely this is the mechanism underlying inhibition by *Metschnikowia pulcherrima*, when this species, by producing pulcherriminic acid, depletes the iron present in the medium, making it unavailable to the other yeasts [51]. During bottling, a sugar addition to smooth mouth-feel up to 10 g/L does not increase susceptibility to spoilage in wines with high ethanol, and so extra doses of sulfur dioxide to compensate higher sugar are not required, which is contrary to common knowledge [52]. However, if cells have the ability to grow, higher residual sugar stimulates proliferation [53].

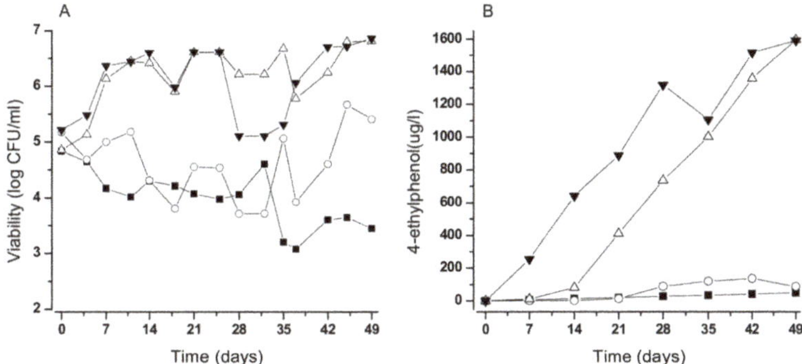

Figure 3. Growth (**A**) and 4-EP production (**B**) by *B. bruxellensis* under wines stored at different temperatures (symbols: ■, 3 °C; ○, 10 °C; △, 15 °C; ▼, 20 °C). Cells were inoculated in 1 L Schott flasks and incubated without agitation to mimic tank storage conditions.

The option to kill actively growing populations through heat treatments using mild temperatures in low volume tanks (35 °C overnight) or pasteurization regimes at bottling can be successful, because Brett is not heat resistant. A solution for contaminated bottled wine is to keep the bottles at 35 °C and check until viable counts fall to 0/bottle.

6.3. Curative Measures

When the wine is off-tainted, there are no effective curative measures without depreciating it. In this situation, we always weigh the possibility of blending tainted wine with "clean" wine. Although this measure may attenuate the defect of the tainted wine by dilution, it cannot be seen as a curative measure. In fact, mixtures of wines free of 4-EP are only effective for small proportions of tainted wines, because large volumes of "clean" wine must be used to obtain a blend with 4-ethylphenol levels lower than the preference threshold.

The effective reduction of 4-EP levels may be obtained by adsorbents (e.g., yeast lees, fining agents, activated charcoal) [54] but wine favorable aroma compounds or color are also removed and a balance must be drawn between the benefits and losses of wine attributes. Our experience with activated charcoal reveals that it may be an efficient option when preventive measures have failed (Figure 4). In this case, besides a reduction in VPs, the wine was also reduced in color and flavor intensity but it was perfectly adequate for blending with un-tainted wines. The use of nanofiltration together with activated charcoal (process accepted by the OIV) and reverse osmosis are other alternatives claimed to be effective in VP reduction. It is also possible that esterified cellulose may be used in filtration sheets to decrease volatile phenols [55].

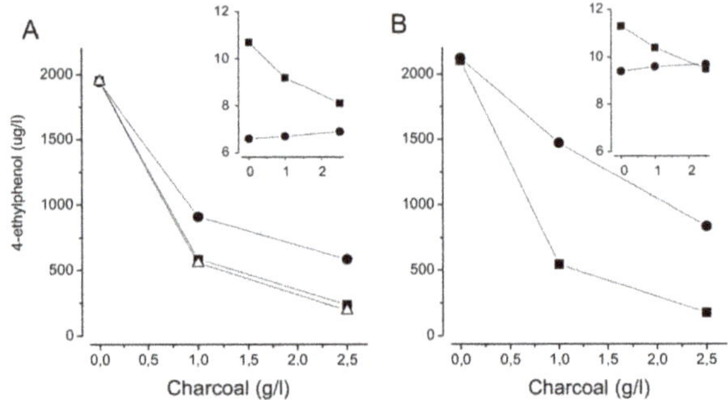

Figure 4. Effect of activated charcoal on the reduction of 4-EP (µg/L) in two red different wines (**A**,**B**) after 3 (●), 10 (■) and 21 days (Δ) of contact. Insertions: effect on color intensity (■) and hue (×10, ●) after 10 days of contact.

7. The Brett Sequel: from "Terroir" to "Terror"

The flavors responsible for the Brett character were certainly known for a long time. It is feasible that it contributed to the so-called "gout du terroir", used to describe certain tainted wines in France by the mid 20th century. Bulk Californian red wines by 1960–1970 were also frequently affected by this taint (Ralph Kunkee, personal communication). In our case, the awareness of the problem begun in the last decade of the last century, when some problematic red wines were tasted by Prof. Dennis Dubourdieu in a workshop organized by our faculty. Coincidently, wine became a fashionable product, worthy of global exposure in the USA. It was a time when noticeably tainted phenolic Bordeaux wines were taken as the expression of terroir by leading wine critics. In addition, barrique utilization increased worldwide to cope with consumer preferences and increasing Brett infection risks. Before winemakers could deal with the problem it was already the object of lively discussions in specialized journals and consumer related websites.

A complex Brett aroma wheel was recently published [56], describing quite variable descriptors, not all unpleasant, such as earthy, leather, savory, spicy or woody, that may be found in high quality red wines. Despite this complexity, the result of generalized "horse sweat" awareness is that, today, every nose seems to be particularly appropriate to find an off-flavor. In fact, the recognition of volatile phenols seems to be equivalent in individuals from very different origins or backgrounds [15]. However, the question is not the recognition of the flavor but the fact that at the least perception of animal notes, the wine may be readily incriminated as an excessively barnyard tainted wine. Particularly in fine wines, bottled for decades, the aging *bouquet* may have notes resembling volatile phenols, albeit being present in low levels. In a world of uniform globalized taste, dominated by the pleasantness of intense fruity-oak flavors and a full-sweet mouthfeel, there seems to be little tolerance to wines with leather-game scents. As a result, winemakers are haunted by the terror of having the slightest note imparted by volatile phenols and become easy victims of technical alternatives that are frequently ineffective to calm their nightmares.

The Issue of Chemical Limits

Given that volatile phenols, where 4-ethylphenol is by far the most relevant, are regarded as spoilage molecules, it is expected that the wine regulatory bodies and large retail companies will try to establish limits for the acceptability of wines based on the average sensory detection threshold of 500 µg/L of 4-ethylphenol. This can be compared with the classic limits for volatile acidity, which is a chemical indicator of spoilage by acetic acid bacteria. However, threshold values for volatile phenols must be regarded only as indicative of horse-sweat taint. Given the different aromatic integration of 4-ethylphenol smell in the overall wine flavor, well-known by wine professionals and a concept explained by Clark Smith in his book Postmodern Winemaking [57], high concentrations are frequently not detected by trained tasters (Figure 5). Therefore, clearly tainted wines can only be determined by sensory analysis. The odorant chemical concentration can only be directly related with Brett activity, as described before.

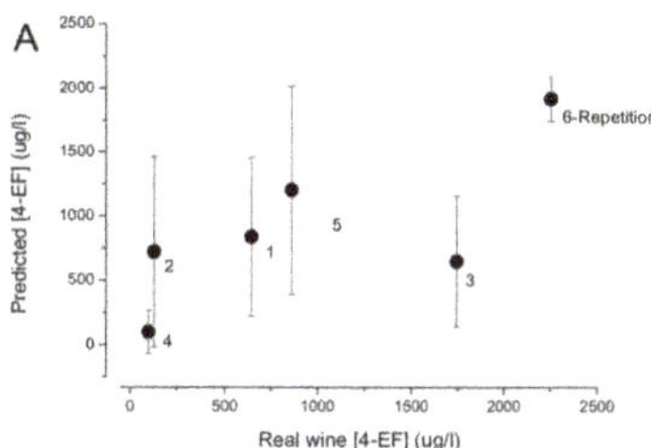

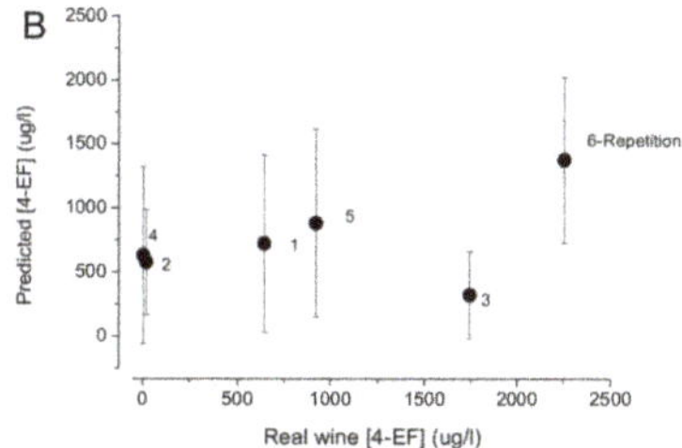

Figure 5. Relation between the real chemical concentration of 4-ethylphenol and the respective predicted concentration given by two different tasting panels ((**A**) DOC certification office; (**B**) faculty enology students). Wines from 1 to 5 are commercial brands. Wine 6 is a blank sample spiked with of 4-ethylphenol (2000 µg/L) and 4-ethylguaiacol (250 µg/L). Vertical bars indicate standard deviation.

8. Final Remarks

According to our experience, the control of *B. bruxellensis* spoilage is, presently, the most serious microbial problem in red wine quality and poses serious constraints to sulfite reduction in its production. There are several technical alternatives that prevent or kill contaminant populations, which contribute to a decrease in its utilization, but they are not fully effective when used alone. Winemakers should learn how to live with these yeasts by monitoring their activity, applying appropriate inactivation measures only when necessary, and weighing the cost-benefit of each option. Besides their technical expertise, winemakers should also communicate to consumers and other wine professionals the difference between volatile phenol sensory detection and wine depreciation, enabling them to understand the diversity and aptitudes of wines with animal-leather flavors.

Acknowledgments: Research was funded by the Fundação para a Ciêcnia e Tecnologia (FCT) trough project UID/AGR/04129/2013.

Conflicts of Interest: The author has no conflict of interest. The founding sponsors had no role in the design of the study; in the collection, analyses, or interpretation of data; in the writing of the manuscript, and in the decision to publish the results.

References

1. Schifferdecker, A.J.; Dashko, S.; Ishchuk, O.P.; Piškur, J. The wine and beer yeast *Dekkera bruxellensis*. *Yeast* **2014**, *31*, 323–332. [CrossRef] [PubMed]
2. Scheffers, W.A.; Wikén, T.O. The Custers effect (negative Pasteur effect) as a diagnostic criterion for the genus *Brettanomyces*. *Antonie Leeuwenhoek* **1969**, *35*, A31–A32.

3. Chatonnet, P.; Dubourdieu, D.; Boidron, J.-N.; Pons, M. The origin of ethylphenols in wines. *J. Sci. Food Agric.* **1992**, *60*, 165–178. [CrossRef]
4. Curtin, C.; Varela, C.; Borneman, A. Harnessing improved understanding of *Brettanomyces bruxellensis* biology to mitigate the risk of wine spoilage. *Aust. J. Grape Wine Res.* **2015**, *21*, 680–692. [CrossRef]
5. Loureiro, V.; Malfeito-Ferreira, M. Spoilage activities of *Dekkera/Brettanomyces* spp. In *Food Spoilage Microorganisms*; Blackburn, C., Ed.; Woodhead Publishers: Cambridge, UK, 2006; Chapter 13; pp. 354–398.
6. Renouf, V.; Lonvaud-Funel, A.; Coulon, J. The origin of *Brettanomyces bruxellensis* in wines: A review. *J. Int. Sci. Vigne Vin* **2007**, *41*, 161–173. [CrossRef]
7. Suárez, R.; Suárez-Lepe, J.A.; Morata, A.; Calderón, F. The production of ethylphenols in wine by yeasts of the genera *Brettanomyces* and *Dekkera*: A review. *Food Chem.* **2007**, *102*, 10–21. [CrossRef]
8. Oelofse, A.; Pretorius, I.S.; du Toit, M. Significance of *Brettanomyces* and *Dekkera* during Winemaking: A Synoptic Review. *S. Afr. J. Enol. Vitic.* **2008**, *29*, 128–144. [CrossRef]
9. Wedral, D.; Shewfelt, R.; Frank, J. The challenge of *Brettanomyces* in wine. *LWT Food Sci. Technol.* **2010**, *43*, 1474–1479. [CrossRef]
10. Zuehlke, J.M.; Petrova, B.; Edwards, C.G. Advances in the control of wine spoilage by *Zygosaccharomyces* and *Dekkera/Brettanomyces*. *Annu. Rev. Food Sci. Technol.* **2013**, *4*, 57–78. [CrossRef] [PubMed]
11. Steensels, J.; Daenen, L.; Malcorps, P.; Derdelinckx, G.; Verachtert, H.; Verstrepen, K. Brettanomyces yeasts—From spoilage organisms to valuable contributors to industrial fermentations. *Int. J. Food Microbiol.* **2015**, *206*, 24–38. [CrossRef] [PubMed]
12. Smith, B.D.; Divol, B. *Brettanomyces bruxellensis*, a survivalist prepared for the wine apocalypse and other beverages. *Food Microbiol.* **2016**, *59*, 161–175. [CrossRef] [PubMed]
13. Agnolucci, M.; Tirelli, A.; Cocolin, L.; Toffanin, A. *Brettanomyces bruxellensis* yeasts: Impact on wine and winemaking. *World J. Microbiol. Biotechnol.* **2017**, *33*, 180. [CrossRef] [PubMed]
14. Goode, J.; Harrop, S. Wine Faults and Their Prevalence: Data from the World's Largest Blind Tasting. In Proceedings of the 20th Entretiens Scientifiques Lallemand, Horsens, Denmark, 15 May 2008.
15. Schumaker, M.; Chandra, M.; Malfeito-Ferreira, M.; Ross, C. Influence of *Brettanomyces* ethylphenols on red wine aroma evaluated by consumers in the United States and Portugal. *Food Res. Int.* **2017**, *100*, 161–167. [CrossRef] [PubMed]
16. Hixson, J.; Hayasaka, Y.; Curtin, C.; Sefton, M.; Taylor, D. Hydroxycinnamoyl Glucose and Tartrate Esters and Their Role in the Formation of Ethylphenols in Wine. *J. Agric. Food Chem.* **2016**, *64*, 9401–9411. [CrossRef] [PubMed]
17. Schopp, L.; Lee, J.; Osborne, J.; Chescheir, S.; Edwards, C. Metabolism of Nonesterified and Esterified Hydroxycinnamic Acids in Red Wines by *Brettanomyces bruxellensis*. *J. Agric. Food Chem.* **2013**, *61*, 11610–11617. [CrossRef] [PubMed]
18. Hixson, J.L.; Sleep, N.R.; Capone, D.L.; Elsey, G.M.; Curtin, C.D.; Sefton, M.A.; Taylor, D.K. Hydroxycinnamic Acid Ethyl Esters as Precursors to Ethylphenols in Wine. *J. Agric. Food Chem.* **2012**, *60*, 2293–2298. [CrossRef] [PubMed]
19. Cabrita, M.J.; Torres, M.; Palma, V.; Alves, E.; Patão, R.; Costa Freitas, A.M. Impact of malolactic fermentation on low molecular weight phenolic compounds. *Talanta* **2008**, *74*, 1281–1286. [CrossRef] [PubMed]
20. Chescheir, S.; Philbin, D.; Osborne, J.P. Impact of *Oenococcus oeni* on wine hydroxycinnamic acids and volatile phenol production by *Brettanomyces bruxellensis*. *Am. J. Enol. Vitic.* **2015**, *66*, 357–362. [CrossRef]
21. Madsen, M.G.; Edwards, N.K.; Petersen, M.A.; Mokwena, L.; Swiegers, J.H.; Arneborg, N. Influence of *Oenococcus oeni* and *Brettanomyces bruxellensis* on hydroxycinnamic acids and volatile phenols of aged wine. *Am. J. Enol. Vitic.* **2017**, *68*, 23–29. [CrossRef]
22. Dias, L.; Pereira-da-Silva, S.; Tavares, M.; Malfeito-Ferreira, M.; Loureiro, V. Factors affecting the production of 4-ethylphenol by the yeast *Dekkera bruxellensis* in enological conditions. *Food Microbiol.* **2003**, *20*, 377–384. [CrossRef]
23. Chandra, M.; Madeira, I.; Coutinho, A.; Albergaria, M.; Malfeito-Ferreira, M. Growth and volatile phenol production by *Brettanomyces bruxellensis* in different grapevine varieties during fermentation and in finished wine. *Eur. Food Res. Technol.* **2015**, *242*, 487–494. [CrossRef]
24. Barata, A.; Caldeira, J.; Botellheiro, R.; Pagliara, D.; Malfeito-Ferreira, M.; Loureiro, V. Survival patterns of *Dekkera bruxellensis* in wines and inhibitory effect of sulphur dioxide. *Int. J. Food Microbiol.* **2008**, *121*, 201–207. [CrossRef] [PubMed]

25. Barata, A.; Correia, P.; Nobre, A.; Malfeito-Ferreira, M.; Loureiro, V. Growth and 4-ethylphenol production by the yeast *Pichia guilliermondii* in grape juices. *Am. J. Enol. Vitic.* **2006**, *57*, 133–138.

26. Dias, L.; Dias, S.; Sancho, T.; Stender, H.; Querol, A.; Malfeito-Ferreira, M.; Loureiro, V. Identification of yeasts isolated from wine related environments and capable of producing 4-ethylphenol. *Food Microbiol.* **2003**, *20*, 567–574. [CrossRef]

27. Martorell, P.; Barata, A.; Malfeito-Ferreira, M.; Fernández-Espinar, M.; Loureiro, V.; Querol, A. Molecular typing of the yeast species *Dekkera bruxellensis* and *Pichia guilliermondii* recovered from wine related sources. *Int. J. Food Microbiol.* **2006**, *106*, 79–84. [CrossRef] [PubMed]

28. Ison, R.; Gutteridge, C. Determination of the carbonation tolerance of yeasts. *Lett. Appl. Microbiol.* **1987**, *5*, 11–13. [CrossRef]

29. Rodrigues, N.; Gonçalves, G.; Pereira-da-Silva, S.; Malfeito-Ferreira, M.; Loureiro, V. Development and use of a new medium to detect yeasts of the genera *Dekkera/Brettanomyces* spp. *J. Appl. Microbiol.* **2001**, *90*, 588–599. [CrossRef] [PubMed]

30. Renouf, V.; Lonvaud-Funel, A. Development of an enrichment medium to detect *Dekkera/Brettanomyces bruxellensis* a spoilage yeast, on the surface of grape berries. *Microbiol. Res.* **2007**, *162*, 154–167. [CrossRef] [PubMed]

31. Guerzoni, M.E.; Marchetti, R. Analysis of yeast flora associated with grape sour rot and of the chemical disease markers. *Appl. Environ. Microbiol.* **1987**, *53*, 571–576. [PubMed]

32. Connell, L.; Stender, H.; Edwards, C.G. Rapid detection and identification of *Brettanomyces* from winery air samples based on peptide nucleic acid analysis. *Am. J. Enol. Vitic.* **2002**, *53*, 322–324.

33. Renouf, V.; Perello, M.-C.; De Revel, G.; Lonvaud-Funel, A. Survival of wine microorganisms in the bottle during storage. *Am. J. Enol. Vitic.* **2007**, *58*, 379–386.

34. Malfeito-Ferreira, M.; Rodrigues, N.; Loureiro, V. The influence of oxygen on the "horse sweat taint" in red wines. *Ital. Food Beverage Technol.* **2001**, *24*, 34–38.

35. Malfeito-Ferreira, M. Yeasts and wine off-flavours: A technological perspective. *Ann. Microbiol.* **2011**, *61*, 95–102. [CrossRef]

36. Capozzi, V.; Di Toro, M.R.; Grieco, F.; Michelotti, V.; Salma, M.; Lamontanara, A.; Russo, P.; Orrù, L.; Alexandre, H.; Spano, G. Viable But Not Culturable (VBNC) state of *Brettanomyces bruxellensis* in wine: New insights on molecular basis of VBNC behaviour using a transcriptomic approach. *Food Microbiol.* **2016**, *59*, 196–204. [CrossRef] [PubMed]

37. Albergaria, H.; Francisco, D.; Gori, K.; Arneborg, N.; Gírio, F. *Saccharomyces cerevisiae* CCMI 885 secretes peptides that inhibit the growth of some non-Saccharomyces wine-related strains. *Appl. Microbiol. Biotechnol.* **2010**, *86*, 965–972. [CrossRef] [PubMed]

38. Oro, L.; Ciani, M.; Bizzaro, D.; Comitini, F. Evaluation of damage induced by Kwkt and Pikt zymocins against *Brettanomyces/Dekkera* spoilage yeast, as compared to sulphur dioxide. *J. Appl. Microbiol.* **2016**, *121*, 207–214. [CrossRef] [PubMed]

39. Costa, A.; Barata, A.; Malfeito-Ferreira, M.; Loureiro, V. Evaluation of the inhibitory effect of dimethyl dicarbonate (DMDC) against wine microorganisms. *Food Microbiol.* **2008**, *25*, 422–427. [CrossRef] [PubMed]

40. Petrova, B.; Cartwright, Z.M.; Edwards, C.G. Effectiveness of chitosan preparations against *Brettanomyces bruxellensis* grown in culture media and red wines. *J. Int. Sci. Vigne Vin* **2016**, *50*, 49–56. [CrossRef]

41. Fabrizio, V.; Vigentini, I.; Parisi, N.; Picozzi, C.; Compagno, C.; Foschino, R. Heat inactivation of wine spoilage yeast *Dekkera bruxellensis* by hot water treatment. *Lett. Appl. Microbiol.* **2015**, *61*, 186–191. [CrossRef] [PubMed]

42. Delsart, C.; Grimi, N.; Boussetta, N.; Miot-Sertier, C.; Ghidossi, R.; Vorobiev, E.; Mietton-Peuchot, M. Impact of pulsed-electric field and high-voltage electrical discharges on red wine microbial stabilization and quality characteristics. *J. Appl. Microbiol.* **2016**, *120*, 152–164. [CrossRef] [PubMed]

43. Van Wyk, S.; Silva, F.V. High pressure processing inactivation of *Brettanomyces bruxellensis* in seven different table wines. *Food Control* **2017**, *81*, 1–8. [CrossRef]

44. Röder, C.; König, H.; Fröhlich, J. Species-specific identification of *Dekkera/Brettanomyces* yeasts by fluorescently labeled DNA probes targeting the 26S rRNA. *FEMS Yeast Res.* **2007**, *7*, 1013–1026.

45. Agnolucci, M.; Scarano, S.; Rea, F.; Toffanin, A.; Nuti, M. Detection of *Dekkera/Brettanomyces bruxellensis* in pressed Sangiovese grapes by real time PCR. *Ital. J. Food Sci.* **2007**, *19*, 153–164.

46. Longin, C.; Julliat, F.; Serpaggi, V.; Maupeu, J.; Bourbon, G.; Rousseaux, S.; Guilloux-Benatier, M.; Alexandre, H. Evaluation of three *Brettanomyces* qPCR commercial kits: Results from an interlaboratory study. *J. Int. Sci. Vigne Vin* **2016**, *50*, 223–230. [CrossRef]

47. Vendrame, M.; Manzano, M.; Comi, G.; Bertrand, J.; Iacumin, L. Use of propidium monoazide for the enumeration of viable *Brettanomyces bruxellensis* in wine and beer by quantitative PCR. *Food Microbiol.* **2014**, *42*, 196–204. [CrossRef] [PubMed]

48. Tristezza, M.; Lourenço, A.; Barata, A.; Brito, L.; Malfeito-Ferreira, M.; Loureiro, V. Susceptibility of wine spoilage yeasts and bacteria in the planktonic state and in biofilms to disinfectants. *Ann. Microbiol.* **2010**, *60*, 549–556. [CrossRef]

49. Barata, A.; Laureano, P.; D'Antuono, I.; Martorell, P.; Stender, H.; Malfeito-Ferreira, M.; Querol, A.; Loureiro, V. Enumeration and identification of 4-ethylphenol producing yeasts recovered from the wood of wine ageing barriques after different sanitation treatments. *J. Food Res.* **2013**, *2*, 140–149. [CrossRef]

50. Umiker, N.L.; Descenzo, R.A.; Lee, J.; Edwards, C.G. Removal of *Brettanomyces bruxellensis* from red wine using membrane filtration. *J. Food Process. Preserv.* **2013**, *37*, 799–805. [CrossRef]

51. Oro, L.; Ciani, M.; Comitini, F. Antimicrobial activity of *Metschnikowia pulcherrima* on wine yeasts. *J. Appl. Microbiol.* **2014**, *116*, 1209–1217. [CrossRef] [PubMed]

52. Chandra, M.; Barata, A.; Ferreira-Dias, S.; Malfeito-Ferreira, M.; Loureiro, V. A response surface methodology study on the role of factors affecting growth and volatile phenol production by *Brettanomyces bruxellensis* ISA 2211 in wine. *Food Microbiol.* **2014**, *42*, 40–46. [CrossRef] [PubMed]

53. Barata, A.; Pagliara, D.; Piccininno, T.; Tarantino, F.; Ciardulli, W.; Malfeito-Ferreira, M.; Loureiro, V. The effect of sugar concentration and temperature on growth and volatile phenol production by *Dekkera bruxellensis* in wine. *FEMS Yeast Res.* **2008**, *8*, 1097–1102. [CrossRef] [PubMed]

54. Milheiro, J.; Filipe-Ribeiro, L.; Cosme, F.; Nunes, F. A simple, cheap and reliable method for control of 4-ethylphenol and4-ethylguaiacol in red wines. Screening of fining agents for reducing volatile phenols levels in red wines. *J. Chromatogr. B* **2017**, *1041–1042*, 183–190. [CrossRef] [PubMed]

55. Larcher, R.; Puecher, C.; Rohregger, S.; Malacarne, M.; Nicolini, G. 4-Ethylphenol and 4-ethylguaiacol depletion in wine using esterified cellulose. *Food Chem.* **2012**, *132*, 2126–2130. [CrossRef]

56. Joseph, L.; Albino, E.; Bisson, L. Creation and Use of a *Brettanomyces* Aroma Wheel. *Catalyst* **2017**, *1*, 12–30.

57. Smith, C. *Postmodern Winemaking, Rethinking the Modern Science of an Ancient Craft*; University California Press: Berkeley, CA, USA, 2013; ISBN 9780520282599.

MDPI

St. Alban-Anlage 66

4052 Basel

Switzerland

Tel. +41 61 683 77 34

Fax +41 61 302 89 18

www.mdpi.com

Beverages Editorial Office

E-mail: beverages@mdpi.com

www.mdpi.com/journal/beverages